This book describes the journey of BrahMos which made India a World Leader in missile technology with the fastest, high precision Supersonic Cruise Missile, realized in short time with a novel collaborative effort between India and Russia. It is the first time experience in an unexplored path to demonstrate "we can do it".

BRAHMOS
THE PATH UNEXPLORED

THE PATH UNEXPLORED

A. SIVATHANU PILLAI

Foreword by
DR. APJ ABDUL KALAM

PENTAGON PRESS

The Path Unexplored

ISBN 978-81-8274-790-6

First Published 2014

Published by

PENTAGON PRESS

206, Peacock Lane, Shahpur Jat,
New Delhi-110049
Phones: 011-64706243, 26491568
Telefax: 011-26490600
Email: rajan@pentagonpress.in
Website: www.pentagonpress.in

Branch:
Prime Arcade
Office #11
1154 Saifee Stret
Opp. M.G.Road, Camp
Pune-411001
Email: pentagonpresspune@gmail.com

Design and layout: Heena Sodhi
Aegean Offset Printers, Greater Noida

Dedication

This book is dedicated to the
numerous contributors from India and Russia for the success;
"Team BrahMos" and
Everlasting friendship between
India & Russian Federation

Royalty accrued to the author through the sale of this book, will be contributed to the Welfare of "Team BrahMos".

Contents

Foreword

Dr. A.P.J. Abdul Kalam
Former President of India

10, Rajaji Marg
New Delhi-110011

A Unique Joint Technological Venture

The Joint Venture 'BrahMos' signifies great friendship and trust between India and Russia in high technology cooperation. The best talent from both nations have been pooled together to achieve the supersonic cruise missile which has outsmarted the developed world.

I was always longing, when India will become the first Nation to possess the most advanced system in the world. BRAHMOS made India proud by making it the first Nation to have a universal operational supersonic cruise missile delivered to the Indian Armed Forces thereby demolishing the Sixth Nation Syndrome. There is no equivalent operational system elsewhere in the world to have speed, precision and power and the universality demonstrated by BRAHMOS. I also know that we started BrahMos with a mere $ 300 million investment by both countries and today the business of the company has crossed $ 6 billion. My desire is now to see the BrahMos business crossing beyond $ 10 billion and to achieve a unique status in the world, connecting the oceans and making new strategic partners beyond the borders.

It is a long story how we could make BrahMos a reality. The path-breaking formation of Joint Venture was not easy. My friend Dr. Yefremov, the then Director General of NPO Mashinostroyenia, is a unique person of respect, who got convinced after visiting the DRDO laboratories that Joint Venture is feasible. This is a unique joint venture with a ratio of 50.5:49.5 giving full empowerment to the JV Company the result of which has traversed its course from mind to market. BrahMos is a fine example of role model of courage and leadership. Different versions, continuous product improvement and integration of the user at every stage made the weapon system user-friendly. All the three wings of the Indian Armed Forces cherish BRAHMOS as their product to be deployed as a prime weapon. Thus, BrahMos is a live example of excellent leadership, system design, system engineering, system integration and system management. The results show that this successful model must be replicated in many other areas both in military and in civil as a national model.

Dr. A Sivathanu Pillai, Distinguished Scientist and Chief Controller Research & Development of DRDO and also the Founder CEO 8s MD of BrahMos Aerospace was my colleague for more than three decades starting from SLV-3 days to IGMDP days and beyond. He made significant contributions in the launch vehicle programme of ISRO and in realizing critical missile technologies for Guided Missile Programme of DRDO through a network of academic institutions and industries thus combating effectively the technology control regime. Dr. Pillai's great leadership in bringing out BRAHMOS from mind to market, integrating the best brains of India and Russia and multiple institutions and industries to realize this world class product is commendable. BRAHMOS missile has been inducted in the Indian Armed Forces to their complete satisfaction.

BrahMos has made a page in the history of the world and Dr. Pillai is synonymous with BrahMos. He has shared many of his interesting experiences during the process of evolution of BrahMos, in this book. I consider this book a valuable narration to the young scientists, technologists, techno-managers and the youth and experienced who aspire to excel in this competitive world. I wish the readers to emulate the role model BrahMos Joint Venture and learn from its experiences.

28February 2014

(A.P.J.Abdul Kalam)

Prolusion

The book is a great creative work dedicated to the history of formation, development and prospects of the most modern forms of military-technical cooperation - a joint Russian-Indian project to develop cruise missile weapon complexes for different types of platforms (Project "BrahMos")

The value addition of this book comes from the fact that the author is a Distinguished Scientist, Managing Director and Chief Executive Officer of the Joint Venture "BrahMos" has been at the forefront and has been actively participated from the initial stages of the formation of this Joint Venture as an irreplaceable leader.

The book highlights the development of weapon complexes right from the design to testing and deliveries of serially produced items which form the entire product life cycle.

All these stages of development are described in the book in detail and in an interesting manner. Considerable attention is paid on the experiences of conducting flight trials of the missiles from various platforms - ground based weapon complex, ship based and underwater platform against different sea and land-based targets. These trials established the high efficiency and reliability of the jointly developed missile. The missile capable of hitting both sea and ground targets has become universal, thus enhancing the combat capabilities of the weapon system.

In the book, special attention is paid to APJ Abdul Kalam and Herbert Efremov, who were the then Director Generals of DRDO and NPO Mashinostroyenia respectively. The idea of creation of the Russian-Indian Joint Venture came from them and they were instrumental in the realisation of the Joint Venture BrahMos.

It would be noteworthy to mention that the author has identified and described in the book, the proposed futuristic developments of the project, including those relating to the enhancement of infrastructure facilities for increased production capabilities, repair and maintenance of equipment and development of futuristic versions of the missile with improved characteristics.

I feel that the proposed book would be very useful not only to specialists in Russia and India, who are related to defence technology but also to the professionals from other countries.

I wish the author of the book and the team headed by him, many more successes and achievements in the implementation of the joint project, which is of great importance for our countries.

A.G. Leonov

Director General
NPO Mashionostroyenia Moscow

Preface

One day, after a phenomenally successful flight test of BRAHMOS from a naval warship in the Arabian Sea, I was on a long journey sailing back towards the western coast of India near Goa. There was a celebration in the ship on the successful launch of BRAHMOS. What a great system we have! It is flawless, the fastest and precise to reach its target every time. I returned to the Captain's Chamber which was allotted to me for the journey and started writing down the missile performance and tasks lying ahead of us. But I thought I would need fresh air to dream further. So, I went to the Bridge, on the top of the ship and sat on the Fleet Commander's high raised chair covered with white cloth. I was alone in the Bridge and I started looking at the beauty of the nature. Light cool sea breeze of the evening was embracing me. The setting sun's scattered light glitter the entire sea with its refracted light and the sun started dipping and the orange coloured rays go above piercing the clouds. Dark blue but almost black sea with its rhythmic waves hitting the surface of the ship and the water droplets fell on the deck. The gorgeous fish jumped out and went inside the water repeatedly. It was now the transition to night. I looked up. The moon was rising and started spreading its shining milky white light. Some stars started sparkling. I was in different world. What a beautiful scene in the Arabian Sea.

Suddenly, my mind stopped enjoying the beauty of the scene. All the past events that had led to this magnificent success of this missile project were running like a flash-back in my mind. It all started with the bull's eye hit on the decommissioned target ship by the missile which traveled with multiple waypoints and made the target burnt with its shear speed and left over fuel. The results of the flight trial were like a textbook performance achieving all the mission parameters and successfully eliminating the target. What a great feeling it gave to our Indian Navy. The unified efforts of the Indian and Russian scientists resulted into robust design of the missile with repeated successes meeting the diverse requirements of the Armed Forces and have stunned the missile scientists of the world with its outstanding performance and reliability. Its high performance capabilities are par excellence when compared to any other cruise missile developed in the rest of the world. I was in an emotional mood that BRAHMOS had proved yet again its superiority to the world.

I traveled many times after successful launches onboard Naval Ships and every time happiness engulfed me with celebrations of the naval crew. But, this time, the flight trial gave me a different mood which made me to recollect the initial days of the project and to think about what led to the success of the joint venture. Those days were the most productive days for the JV Company since they formed the foundation of this mighty project. The countless man-hours we spent gave a new dimension to the Joint Venture.

In 1995, when the idea was crystallized to form a JV, I was the Program Director for Integrated Guided Missile Development Programme (IGMDP) and I was asked to be the CEO&MD of the JV Company by Dr. Kalam with the approval of Government of India. Since I was the strong proponent for the entry of cruise missile in India and had worked for the formation of JV from the beginning, I was the natural choice as the CEO&MD of the Company in addition to be the Chief Controller (Research and Development) for all the Missiles and the Range, at DRDO HQ, New Delhi, concurrently holding the position of CEO&MD for the joint venture.

Many of my well wishers including senior bureaucrats were of the opinion that JV with Russia cannot be a success and told me not to take the responsibility. The main reason for their view was that the Soviet Union had broken and the economic situation in Russia was declining rapidly during 1990s. Moreover, the experiences of earlier JVs were not encouraging, especially in advanced technology areas, as the risk was high in product realization. And above all, there was no Quality Requirement from the Services. But I was convinced that whatever I have learnt from the great visionaries - the 'men of honour' who laid the foundation for the technologically developed India - Dr. Vikram Sarabhai, Prof. Satish Dhawan, and Dr. Abdul Kalam would stand with me by my side to ensure that we achieve success in our mission. The very thought about them gave me the determination to face the challenges that might come. Their way of encouraging even small achievements of young engineers and scientists regardless of their positions and their soft way of condemning wrong deeds are legendary. I feel that more have to be written about those great men who dedicated their lives for the technological uplift of India. Success of the Space Programme of ISRO and the missile projects of DRDO bear the testimony of what these men have seeded. If so, then why couldn't I succeed as CEO&MD of the Joint Venture, though it was supposed to be a difficult task?

Those visionary leaders inculcated a wonderful culture that could bring the desired results from nowhere, because talented brains are omnipresent with hard work and perseverance to give efficient output. But to extract the work efficiently, those brains have to be allowed to work freely in the right direction by creative leadership. The political will and the support, experiences of efficient practices of IGMDP, dedicated teams, the best utilization of the great scientific minds of the two countries and cultural fusion, friendship and faith could lead to success of BrahMos.

With strong will power, I took up the charge as CEO&MD of the JV Company BrahMos. Each and every day used to buzz with lot of activities. Since I was holding dual posts, it demanded me to carry out duties as both Chief Controller at DRDO and as CEO&MD at BrahMos which made my days hectic. But the duties of the two positions had to be carried out without conflict of interest. Whenever I think about those days, I get goose bumps and cherish the memories of those fruitful days, ever in our team's minds.

Today, the BrahMos joint venture between India and Russia is very much acclaimed all over the world as the most successful international defence cooperation. The sharing of expertise between the two countries in various missile technologies has made the missile the most powerful in the world in its class. Scientists from India and Russia worked hand in hand from the design phase to the production phase to realize the missile with "Mind to Market" concept. The experiences that we have gained in the design, development, production, marketing stages are tremendous and mind boggling. Today many countries are requesting India to sell BRAHMOS missile systems to them so that they can induct it in their armed forces. At the same time many other countries including those powerful nations are worried. BRAHMOS Supersonic Cruise Missile has been a great technological breakthrough in the Indian Missile programme, thanks to the new format of the joint venture.

Going back in time we see that India had excelled in the past in fields of Mathematics, Science, Engineering, Technology, Medicine, Astronomy, Cosmology and many more. Lot of inventions were made and many of them were very exotic. But unfortunately there were no proper documentations made, perhaps to avoid the replication of those achievements for evil purposes. Many engineering feats that were achieved in the early days are technologically at par with the modern standards, e.g., a dam at South India. This dam is at Tiruchirappalli district of Tamil Nadu state and was built by Karikaal Chola, 2000 years ago. The dam was constructed as a barrage with just stones and mortar, resembles a snake's trail from aerial view. Until the dam was reengineered by the British in the late 18th century it was storing water for more than a millennium. Today the water flow has decreased due to the number of dams across Cauveri and her tributaries in Karnataka and Tamilnadu. But without these dams in those days, the water flow would have been enormous and how such huge amount of water flow was diverted, how the dam was constructed is still not understood by experts. An extraordinary human and animal effort would have been put in but we do not have any document of how such an extraordinary feat in civil engineering was achieved. Similarly the iron pillar at Qutub Minar standing for more than two millennia (912 BC) without rusting is another feat in metallurgy. There is no document available to establish the composition of the pillar and how it was made.

Medical science was also at its zenith because the scientists of ancient India like Acharya Sushruth have conducted even plastic surgeries at that time. Our Siddhas were profound in handling poisonous chemicals like mercury and arsenic to treat dreaded diseases such as cancer. They have used the chemicals in a nano powder form (Pashpam), a proof that nanotechnology is not new to India. During the preparation of such medicines, managing exhausts of poisonous fumes was quite common occurrence because inhaling them would result in death. To overcome such conditions, they used their Yoga skills and controlled their breathings.

Some of the chemical combinations, which they have stated in their scarcely available literatures, are found to be impossible to achieve even in today's modern era. There are records that they have cured some terribly complicated diseases but we do not have any manuals or documentations written by them. I have seen this personally with my own eyes, when I was a ten year old boy, my father Mr. Apathu Katha Pillai (meaning one who saves others from dangers), an expert in Siddha, Ayurveda, and Yoga, who had cured dreadful diseases like cancer, true to his name. But like others he did not leave any record.

There were so many technological marvels, yet there are no documents regarding the process followed. Undoubtedly, India was dominating the world with its vast expanse of knowledge. But in the last one thousand years, the foreign invasions resulted in India plunging into her dark ages of slavery. Her enormous wealth was systematically looted by many invaders. Today, we have come a long way from the curse of slavery. We got freedom because of the sacrifice of the many, including the great leaders. But there is no war memorial at the capital to celebrate the freedom and how we got it and also to honour those great men. Today, our youth is full of enthusiasm to see a prosperous India. But they lack knowledge about the freedom movement and the greatness of our nation in the yester-years. Records should be maintained to help them to develop a sense of patriotism. Records preserve knowledge, and knowledge will help the future generations.

Coming back to the present, BRAHMOS is no small achievement. Everything was started from scratch. Just from a scope of an anti-ship missile development we have evolved a universal missile. It is now a Multi-Platform, Multi-Target, Multi-Role and Multi-Trajectory capable missile. In record time of just less than a decade the missile has emerged from the drawing boards to its deployment by the frontline forces. Certainly, BRAHMOS has made a page in the world history for its power and uniqueness.

When pondering over all these issues, the Captain informed me that the ship is about to reach the shore. After I reached the naval base, a strong thought came to my mind that a book should be written on the BrahMos experiences. If not written properly, our efforts and techniques will be forgotten in the future and the youth have to reinvent what we have invented. So I thought about writing a book on the story of India's transitional voyage - BRAHMOS, whose equivalent is yet to be born.

Because the path that we have travelled was not at all smooth; lot of blood and sweat have been shed to achieve the present level of success. Our achievements have resulted into an integrated defence company in India, taking care of design, development, production, marketing and product support. Many of our methodologies were daring and never before attempted. This book will be of great help for the future leaders, who are aspiring to leap-frog the hurdles and constraints with courage, and make India a global leader.

Since the objective of the book is to reach everyone, I have not gone deep into technology intricacies and at the same time not revealing the secrecy which relates to the design and certain performances as this is a frontline weapon system for our armed forces. The Appendix contains the fundamentals required for understanding a missile. If the reader is a novice to missile terminologies, I suggest the reader to understand the basics of guided missile from the Appendix. This book has been divided into many parts to cover various aspects of the voyage.

Since, the idea of writing book was conceived while I was travelling in the ship, I thought of building up the experiences in different chapters as a voyage on the sea.

Let us begin our voyage through the exciting unexplored path of BrahMos.

Acknowledgements

The top leaders of the Governments of India and Russian Federation gave outstanding support, constant encouragement and continuous motivation through their visits made the joint venture a successful one. Visits of Presidents, Prime Ministers and the Defence Ministers of both the countries had given significant thrust to the Joint Venture and have highly motivated "Team BrahMos". BrahMos is always indebted to all these leaders. Our mentors and visionaries Dr. APJ Abdul Kalam and Dr. HA Yefremov with constant interaction, guidance and advice helped to put the programme on fast track. National Security Advisor Shri Shiv Shankar Menon, Scientific Advisors to Defence Minister of India, Directors of DRDO laboratories, Ambassadors, Secretaries to Government, Indian industry captains, Ambassadors of Indian and Russian Embassies, Officials from Federal Service Military Technical Cooperation from Russia have been the source of encouragement. Above all, the Chiefs of Staff of the three wings of Indian Armed Forces gave outstanding support and demanded higher level of performance which led to continuous upgradation. Dr. AG Leonov, Dr. AA Dergachev, Mr. AN Semaev, Dr. AV Khromouchkin, Mr. AN Strakhov, Mr. AB Maksichev, Mr. VM Kiselev, Mr. VA Merkulov and many scientists from NPOM and partner Russian Industries gave their all out support to the programme to succeed.

'Team BrahMos' consisting of scientists, engineers, retired armed force officers and bureaucrats, supervisors, technicians and staff from DRDO, NPOM and BrahMos have displayed dedication and hard work to shape the joint venture on the right path with excellence in performance. They shared their experiences at every moment of the growth trajectory of BrahMos, making BrahMos a remarkable corner stone of Indo-Russian Joint Venture as a shining example of international cooperation.

While writing this book, many of my colleagues shared their experiences with me. A number of industry meets, partnership programmes and flight tests from ships and shores and interaction with administrators and Service personnel formed as input for the book. I am thankful to all my colleagues and partners who contributed in the evolution of the book, which is written on their behalf.

I would like to specifically acknowledge the great innovative thoughts of Shri B. Kamalanath who has contributed in the preparation of the book. His analysis on the Gulf war is remarkable. I am also thankful to Shri K. Hariharan for his painstaking efforts and continuous upgradation of the text. Both these two colleagues have worked many days and nights with me like the voyage which has been described in the book, to bring it to shape and read worthy. I would also like to thank Mr. Prasannan for his expert views on the contents.

I also thank my colleagues Shri Praveen Pathak, Shri Rohan Mishra, Shri Rupesh Verma and Shri Rajan Arya of Pentagon Press for their help.

Dr. A. Sivathanu Pillai

PART 1

Embarking on a Voyage

"To be secure on Land, we must be Supreme at Sea."

- Pandit Jawaharlal Nehru
First Prime Minister of India

In this part, I will narrate two important incidents which occurred in the Indian Naval History – a moment of Jubilance and a mortal threat. The events made the Indian Navy to hunt for a superior weapon system. But that weapon system was nowhere to be seen. To possess such a weapon system it was understood that an arduous voyage was needed. The voyage was not just to be of Indian Navy's, but it was to be a voyage undertaken by the Nation...

1

A BATTLE AT HIGH SEAS

NIGHT OF 4TH OF DECEMBER 1971

Darkness had thrown its blanket on the buzzing Karachi harbour. Karachi, the headquarters of Pakistan Navy had been an important naval base right from the colonial era. Movement of Merchant vessel was restricted and major portion of that area was under the grip of wartime darkness, because, Pakistan had started a war with India. Pakistan had mounted unsuccessful pre-emptive air strikes to neutralize Indian Air Force (IAF) bases on the previous evening. Due to that, Pakistan was expecting an Indian retaliation from the air, virtually all over Pakistan.

At some time close to midnight, 'PNS Khaibar', a destroyer of Pakistan Navy patrolling that area, observed a bright light in the air approaching her. Since an aerial threat was expected, that bright light was assumed as an attacking aircraft and the Pakistani sailors blazed away the vessel's anti-aircraft guns to shoot down the light. But that undeterred mysterious light kept coming closer and instead of dropping bombs, it rammed the vessel and exploded. Then, like a spike struck in the spine, the truth emerged in the Pakistani minds that the light was not an aircraft, but a missile. This was the first missile fired in anger in the Indian sub-continent. Before the Pakistanis recovered from their surprise and horror, a volley of missiles fired by three Indian warships attacked many Pakistani ships and set harbour installations ablaze, celebrating a sort of Diwali (an Indian festival of lights). But they were merely bigger than fishing trawlers, and technically they were to be called as missile boats. Indian Navy took pains in taking these small vessels near the enemy land and hitting him hard with Soviet 'SS-N-2B Styx' missiles and the rest is history.

THE SUPERPOWER THREAT

Indian Navy faced a grave threat from a superpower in the same war. The task assigned to the Indian Navy was to enforce a naval blockade to prevent vessel movements in both West and East Pakistan. The attack on Karachi was a part of this blockade enforcement. In the Bay of Bengal, the blockade was enforced through the task force centred on the aircraft carrier 'INS Vikrant'. To intimidate India, a task force from the US Navy's Seventh Fleet was dispatched towards Bangladesh in support of Pakistan. The task force was headed by the nuclear powered aircraft carrier 'USS Enterprise', the largest aircraft carrier in the world during that time.

The taskforce consisted of another small carrier carrying twenty five assault helicopters and troops. These vessels were escorted by three warships armed with missiles, a nuclear attack submarine and four other vessels. This force was assigned to break the blockade that had been imposed by 'INS Vikrant' over the Bay of Bengal. The taskforce of 'INS Vikrant' was no match to the American taskforce and was under a great threat. But the much hyped and awaited taskforce didn't come into the scene. Later it was known that the Americans turned away after finding a powerful presence of Soviet Nuclear Submarines in the Indian Ocean in support of India. Thus, the great threat was thwarted. Thanks to the true friendship between India and Soviet Union.

These two important incidents - a triumph and a threat - occurred in the 1971 Indo-Pakistan War, made Indian Navy to think of a superior weapon system capable of engaging threats from longer range. But this search was on for many-many years.

34 YEARS LATER – 15TH OF APRIL 2005

In the Arabian Sea, one of the Indian Navy's frontline warships 'INS Rajput' was patrolling near Mumbai. The morning weather was pleasant and beautiful. But suddenly at 0745 hrs the radar operator shouted, 'enemy vessel!' The radar operator gave the bearing and range of the enemy vessel that was drifting towards Mumbai. The captain of 'INS Rajput' was informed. The captain gave the necessary orders; bells started ringing and speakers screeched orders for the sailors to man their respective stations. Within minutes, the vessel was brought to a wartime state of alertness with all the weapons being ready to fire. Everybody's heart rate was thumping, because this could be a prelude for another war. But the men of Indian Navy held their nerve due to their superior training. Anxious moments of waiting for the enemy to reverse his course were passing away. But the enemy didn't turn, instead he was advancing.

"Fire"! The captain ordered and in no time a missile took off from the deck of INS Rajput. The missile created a thunderous boom in the calm sea and it flew with a speed almost three times greater than the speed of sound. Like a thunderbolt, the missile struck the target vessel, which sank in minutes in two halves as cinematographed in Hollywood movies. With this engagement, a new era dawned in the theatres of war.

Launch from INS Rajput

Target Ship Ex INS Sindhudurg sinking

That missile is the world's best supersonic cruise missile "BRAHMOS". Yes, the incident that I have described was a combat firing drill of BRAHMOS missile and the vessel that was sunk was a decommissioned vessel of the Indian Navy ex 'INS Sindhudurg'. The entire drill was witnessed by the top echelons of Navy including Adm. Arun Prakash, the then Chief of Naval Staff.

In combat, the enemy cannot even think of shooting down BRAHMOS. And what's more? He will never hear the thunder of the missile, because, the missile is supersonic, stealthy, and the sound is minimum while approaching the target. By the time the missile reaches him, he will not be in a position to hear it, because he would have ceased to exist.

When the outcome of the combat firing drill was released to media, the bothering time started for our adversaries. Because, in the event of war, their warships may share the fate which ex 'INS Sindhudurg' had to face. Today, the BRAHMOS supersonic cruise missile is progressively being inducted in the armed forces of India with constantly increased performance capabilities. The missile system was first inducted into the Indian Navy, then into the Indian Army followed by the Indian Air Force. The missile has become a frontline force multiplier weapon system, in the Indian Armed Forces.

Today, with BRAHMOS, unlike in 1971 war, the Indian Navy without moving closer to the enemy territory, can fire the missiles well away from his reach yet inflicting higher level of fatal damage. Plus, Indian Navy does not need the support of an external power to counter superior threats. If an occasion arises of a major fleet (a group of ships) coming to the Indian waters to threaten India, it will be difficult for them to return when Indian naval fleet with BRAHMOS is in operation. There is no other weapon system seen on the horizon, to match and counter BRAHMOS, even after over a decade of our first flight trial on 12 June 2001. Countries who have not reached to deploy a supersonic cruise missile, are dreaming to intercept BRAHMOS when deployed. I wish better luck for them. That does not mean we have to be complacent; Technology always moves ahead. Hypersonics is the next, and so on. It is always necessary to think ahead and develop newer technologies which will keep up the leadership and edge in the war theatre.

PART 2

Unfurling the Sails

> "What all prudent princes* ought to do, who have to regard not only present troubles, but also future ones, for which they must prepare with every energy, because, when foreseen, it is easy to remedy them; but if you wait until they approach, the medicine is no longer in time because the malady has become incurable."
>
> **-Niccolo Machiavelli**

** Prince = Country*

In this part, let us go through an insight as to why India, a nation epitomised as a peace pigeon by the other countries, and respected as an advocate of world peace, required to embark on the development of nuclear weapons and delivery systems (missiles), how they were developed, and then how they formed the foundation for BRAHMOS - The commencement of the voyage...

2

TALONS OF A PIGEON—THE NUCLEAR INDIA

The reader must know why India speaks about disarmament and at the same time develops deadly nuclear missiles and possesses nuclear warheads. They must know that why nuclear India is a country with sufficient maturity, ethical conscience and responsibility.

CONCERNS OF INDIA

What are the types of threats that a nation could face? It could be one among or all of the following threats: economic threats, military threats, and existential threats. Economic threats are the threats posed against a country's economy and are not in purview of this book. Military threats severely undermine the military strength of a nation. Once the military is undermined, the nation is forced to be at the negotiation table, negotiating unfavourable terms. Because, in the words of the most quoted strategist *'Karl von Clausewitz'*, *"War is the continuation of policy (politics) by other means."* It means that war is not merely an act of policy but truly a political instrument - a continuation of political activity by other (military) means. After WW-I, Germany was made to cede territories and pay penalties through a *'Versailles Treaty'* doctored by the allied nations for beginning the war. Generally, economy remains the driving force for conflicts and when favourable economical conditions are achieved by the conflict initiator, the conflict ends. In the history, many wars had their roots to the economy. But there is also a possibility that military threat can originate from racial or religious hatred.

If the motivation is hatred, the enemy after destroying the nation's military, may try to annihilate the population or occupy the nation itself. At this point the military threat transforms into an existential threat for a nation. For example, Israel faces existential threat from its surrounding nations. Since Israel is militarily strong, and it has the massive support of a superpower, it is able to blunt the threats and survives. According to various eyes, if suppose, Israel loses its military might, then an existential threat for Israel will be imminent. In that aspect, what are the threats that India faces?

For India, presently there are no threats, but there are concerns which could become threats. But, if we are sufficiently strong to handle any threats, then peace will continue to prevail and concerns will remain as concerns. What are those concerns? There are two: One is from the immediate neighbourhood and the other in the vicinity. The concern in the immediate neighbourhood is the military prowess and territorial claims of the neighbouring two countries that fought wars with India. In the vicinity we see activities of proliferation of weapons and large scale modernisation of defence equipment. These proliferations could become threat to India in future. As Niccolo Machiavelli says, *"identification of threats at right time is important for future security."* A comprehensive analysis on the threat perceptions and preparedness is essential for the defence of the country.

IN THE NEIGHBOURHOOD

Chinese Territorial Claims

The root cause of troubles in India-China relation is 'British India's and later India's perception towards Tibet, and China's perception towards Tibet. During the times of British India, the boundary between Tibet and India was marked and maintained through a border agreement. The agreement was driven with a basis of pre-independence view and later Indian view that Tibet is an independent country and served as a buffer state. But the view of Communist China is that Tibet is a part of Chinese territory. So China never accepts the demarcated border and says that it is inside their territory.

China occupied Tibet by force in 1950. A huge transgression was made by China into Indian Territory – Aksai Chin. Chinese leaders assured friendly relationship with India during the late 1950s. But the Chinese army invaded India and occupied many Indian positions in a brief border war during 1962. Still China possesses 'Aksai Chin,' a huge chunk of Indian land spreading hundreds of square kilometres in land area. Not only that, China is claiming our state 'Arunachal Pradesh' as theirs and also that Arunachal Pradesh was part of Tibet. China even opposed the visit of Indian Defence Minister there.

Confidence building measures are going on between India and

China, Military to Military contacts exist and also joint exercises are planned with. In the civilian sector, massive amount of trade also exists between India and China.

But militarily, China almost has three times greater military might than India in terms of combat naval vessels, submarines, combat aircraft, artilleries, tanks, and individual soldiers. China has a very vast nuclear weapon stockpile and stands third after Russia and USA. Not only nuclear weapons, China has an array of nuclear capable ballistic and cruise missiles. China also has Submarine Launched Ballistic Missile (SLBM) capabilities. Chinese missiles also have Multiple Independent Re-entry Vehicle (MIRV) warheads, meaning that the missile's warhead will jettison many sub-munitions when it re-enters the atmosphere. Each sub-munition is also a nuclear warhead that can destroy a separate city. A single missile will be sufficient to pepper many enemy cities. Still efforts are continuously on in China to perfect ballistic missile technology to take it to a step further.

Inimical Attitude of Pakistan

Immediately after the independence, Pakistan keeping tribal proxies in the front, invaded Kashmir. Today a major portion of Kashmir is still under Pakistan as what is called as 'Pakistan Occupied Kashmir'. Further Pakistan fought two full blown wars during 1965 and 1971 and a misadventure in Himalayas during 1999, in what is called as 'Kargil War'.

The military might of Pakistan can be levelled to a third of India's military might. This levelling is for conventional forces only, i.e., combat naval vessels, combat aircraft, artilleries, tanks, etc. But in the unconventional front, Pakistan has nuclear capable ballistic missiles to hit India at a place where it can hurt more. Not only the nuclear capable ballistic missiles, but Pakistan is also aggressively developing nuclear capable air launched land attack cruise missiles. Pakistan has a very active nuclear program and reported to possess some 70–110 nuclear warheads. By compromising the economical development, Pakistan has developed these arsenals. Internationally it is feared that deteriorating internal situation of Pakistan may lead to the terrorists getting the control of nuclear weapons and missiles. In that case, the repercussions will be highly unpredictable and gravely serious.

Also the relationship between China and Pakistan is interesting. China had extended support for Pakistan during 1965 and 1971 wars. The Pakistani nuclear programme benefited very much from Chinese assistance. Not only nuclear weapons development program, the missile development program also enjoyed Chinese assistance either covertly or overtly. Many of the Chinese ballistic missiles re appear with a Pakistani name and colours. What the Pakistanis describe their friendship with China is an all weather friendship higher than sky and deeper than ocean.

In India some feverish strategists claim that a two-front war scenario is on the rise. According to them, if Pakistan starts a war with India, then China may strike on the other border also. Also if China starts a war with India, then Pakistan may also attack India. They also add that Pakistan's proxy war capability through its so far nurtured terrorist organisations may create additional troubles along with the Maoist troubles.

It should be noted that if these claims become true then India may have to fight from the marshes of Gujarat, deserts of Rajasthan, icy Himalayan heights, very limitedly accessible North-eastern borders, on the seas, under the seas, cyber space, and also in the outer space – not just with one country but with two. Though outer space is not supposed to be militarised but with China proving their satellite attack capability, there is every possibility that China may remove Indian Satellites from their orbit, in the event of war. Historically two front wars have always been costly and deadly. It is worth to mention that during WW-II when Germany was caught in a two front war with USSR in the east and the other allies in the west, its defeat became a certainty.

We do not know whether India may have to fight a two front war with two nations armed with nuclear tipped ballistic and cruise missiles. But if such a situation occurs then it will be the worst threat that India had ever faced in its history.

CONCERNS IN THE VICINITY: PROLIFERATION OF WEAPONS OF MASS DESTRUCTION

Weapons of Mass Destruction (WMD) are Nuclear, Biological, and Chemical (NBC) weapons that can bring significant harm to humans and can damage the man-made structure. Ballistic missiles and to an extent cruise missiles are used as the delivery systems for these killers. These missiles take these warheads to the enemy territory to kill thousands in a go. A nation wishing to possess a dreadful WMD arsenal needs both delivery systems and the NBC weapons. But first what are these NBC weapons.

NBC Weapons

A conventional warhead is packed with high explosive materials to cause destruction in a limited scale. Effectiveness of these warheads may be only for a few 100 metres. But NBC warheads are meant for mass destructions - for killing civilians on an unimaginable scale. NBC weapons are called as unconventional (warheads) weapons. You know that Science is broadly classified into Physics, Chemistry, and Biology disciplines. Each discipline has one unconventional warhead: Nuclear ones for Physics, Chemical ones for Chemistry and Biological ones for Biology.

Biological warheads contain harmful micro organisms like Anthrax and can contaminate the target area, say a city with these organisms. Such organisms multiply rapidly to cause highly contagious diseases thus causing huge chaos and finally destruction. Imagine the chaos resultant of an entire city's population infected with these warheads and micro organisms waiting to infect anyone who would enter the city. 100 kg of Anthrax spores can kill 42,000 – 1,40,000 people. USA possessed a huge arsenal of biological weapons and has declared that it had destroyed those stockpiles. India has no biological weapons and is a signatory to the Biological Weapon Convention that bans proliferation of Biological weapons.

On the other hand, Chemical weapons release harmful chemicals. There are various types of chemicals like nerve agents that can choke the nervous systems; some damage the respiratory systems; some blind the eyes; some can even lead to the decomposition of flesh. 1000 kg of popular nerve agent 'sarin' when dropped from air can kill 400-800 people. India had chemical weapon stockpile and has destroyed them completely. India is also a signatory of Chemical Weapon Convention that bans the proliferation of chemical weapons.

In the hindsight, a nation with well established bio-chemical technology base can develop both chemical and biological weapons. Expertise will be required in developing delivery systems and disperse mechanisms to disperse these agents.

Whereas, the case of a nuclear weapon is different! A huge infrastructure and a fair amount of expertise are required for developing nuclear weapons. Theoretically, any nation that has an active nuclear programme for power generation can develop a nuclear weapon. How? Nuclear reactors using light waters as moderators need a specific uranium called enriched uranium, a highly radioactive isotope of natural uranium. This isotope exists in very very small quantities in the natural uranium and has to be extracted. This process is called as uranium enrichment. This very same enriched uranium can also be used for nuclear weapons. The key technology in nuclear weapons is making this enriched uranium to become radioactive at a desired time and desired place. To make the uranium radioactive, it must be compressed very hard. Usually, it is done by explosive triggers. This technology development and uranium enrichment facilities can lead to nuclear weapon developments, or at least theoretically.

So far USA, USSR/Russia, UK, France and China have nuclear weapons and are the declared nuclear nations. They are called P5 nations and are permanent members of the UN Security Council. India attained nuclear capability in the year 1974 and became a self declared nuclear weapon state in the year 1998. Pakistan also became a nuclear nation by 1998. There is an interesting case of Israel. Israel is suspected to have nuclear weapons and that nation has neither acknowledged it nor denied. So its nuclear status is not declared. North Korea and Iran are also in that status. There are few other countries who had nuclear capability, but terminated in 1970s. Today, Japan is just a step away from nuclear weapon development. Since Japan has all the technologies and expertise to build a nuclear weapon, it can build quickly, if it requires. But Japan's constitution prevents Japan building of Nuclear Weapons.

When a nation is possessing ballistic missiles either home grown or proliferated or imported, it must be considered as a 'concern' even if that nation is friendly. Not only that nation can modify the missile, but there is also a possibility of sinister condition called 'the missile alliance'.

As long as missiles remain the frontline fighting teeth and as the last line of defence, such missile alliances will keep on emerging. All such alliances have to be kept under a watchful eye. Reports say that Pakistan by transferring nuclear technology to North Korea, received Nodong missiles. Similarly, Iran's "Shahab" missiles have a distinct North Korean signature on them. Thus, every missile 'mass produced' or 'bought' or 'proliferated' should be considered a 'Threat'.

What Do We Infer?

The P5 countries have large number of nuclear weapons and carriers including ICBMs. The availability of such capability makes them powerful to dominate the world affairs. They are the members of the UN Security Council. There are countries like India, Pakistan, Iran, Israel, North Korea who have the nuclear weapons and Intermediate Range Ballistic Missiles (IRBM). There are 24 other countries that have short range ballistic missiles of less than 1000 km range. The data is based on the study by Carnegie Endowment for International

Peace. Today, India is surrounded by many countries that possess destructive weapons and have missile capabilities through imports and/or indigenous development. Particularly, China and Pakistan who have fought wars with India possess nuclear weapons and long range strike capability. The inference is that India cannot lose guard under any pretext since the geopolitical tide may change at any time translating proliferation activities to full-fledged threats.

DEALING THE THREATS: INDIA'S ARMAMENT PROGRAM

Going Nuclear

India going for nuclear weapons is itself a long story. India attained

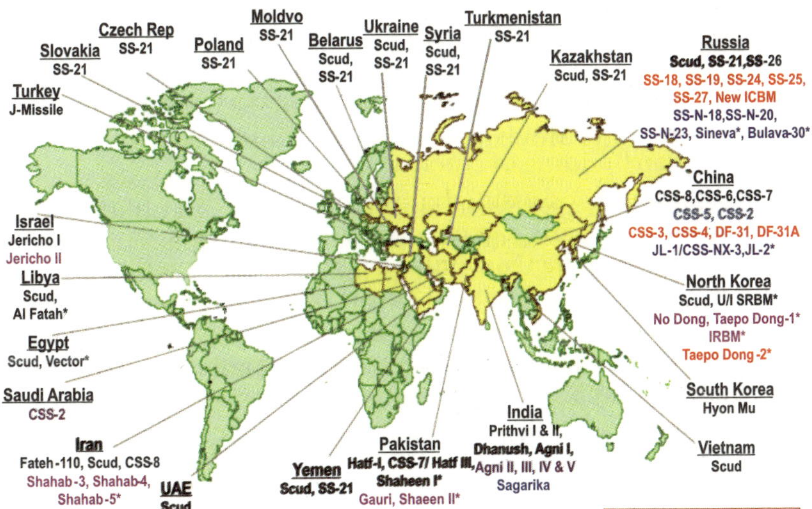

24 Countries have ballistic missiles with ranges < 1,000 km

SRBM
MRBM/IRBM
ICBM
SLBM
* System Under Development

Deployed ICBMs of Declared Nuclear-Weapon States

Country	Missile	Range (km)
China	DF-5	13,000
France	M45 SLBM	6,000
	M4 SLBM	6,000
Russia	SS-18	11,000
	SS-19	10,000
	SS-24	10,000
	SS-25	10,500
	SS-27	10,500
	SS-N-18SLBM	6,500/8,000
	SS-N-20 SLBM	8,300
	SS-N-23 SLBM	8,300
U.K	Trident II/D-5 SLBM	7,400
U.S	Minuteman II	9,650
	MX Peacekeeper	9,650
	Trident I/C-4 SLBM	7,400
	Trident II/D-5 SLBM	7,400

Missiles with ranges exceeding 1,000 km in 6 Countries of Proliferation Concern

Country	Missile Name	Range (km)
India	Agni II	2,000
	Agni V	5,000
Iran	Shahab III	1,300
Israel	Jericho II	1,500
North Korea	No Dong	1,300
	Taepo Dong I	2,000
	Taepo Dong II	5,500
Pakistan	Ghauri I/No Dong	1,300
	Ghauri II	2,000
Saudi Arabia	CSS-2	2,600

Carnegie Endowment for International Peace *Deadly Arsenals* (2002), **www.ceip.org**

Ballistic Missile Environment - Carnegie endowment study for world peace

independence from a foreign occupation through Ahimsa Dharma. In other nations, independence came with a terrible price. Many nations attained freedom after a bloodbath. But never in the history of mankind, had an entire nation stood up against the aggressors without weapons. Their weapon was very new and was coined as "Satyagraha".

Pandit Jawaharlal Nehru, one of the leaders in India's freedom movement, became the first Prime Minister of independent India. Under his visionary leadership, India took centre stage in the world peace initiatives and disarmament of mass destruction weapons, right after India's independence from the British clutches. He spearheaded the 'Non Alignment Movement' as a counter weight to the two superpowers.

He evolved a non-aggressive and tolerant foreign policy expressing brotherhood and co-existence with peace. The policy framework was called as "Panchsheel" - the five principles of peaceful co existence. These five principles were to govern the relationships between the new nations emerging after the colonial era. It was codified formally through a treaty form between China and India in 1954. Those five principles are:

1. Mutual respect for each other's territorial integrity and sovereignty,
2. Mutual non-aggression,
3. Mutual non-interference in each other's internal affairs,
4. Equality and mutual benefit, and
5. Peaceful co-existence.

In Nehru's words, "If these principles were recognized in the mutual relations of all countries, then indeed there would hardly be any conflict and certainly no war." Nehru believed in world peace and in the world arena India started to be seen as a peace pigeon.

In spite of his good efforts, war was forced upon India in 1962. Following this war, China became nuclear in the year 1964. They also continued the development of multiple missile development in a fast track manner. And by mid sixties their very first ballistic missiles, the DF-1 and DF-2 became operational.

During this period, India was in an enviable position in nuclear technology. This was achieved by the pioneering efforts of 'Dr. Homi Jehangir Bhabha', who literally marshaled Indian nuclear energy efforts to the forefront. India's civilian nuclear energy efforts had the military aspect kept open. When Dr. Bhabha died in an air crash in the year 1966, his responsibility was passed on to the hands of Dr. Vikram Sarabhai. After the sudden demise of Dr. Sarabhai in December 1971, Dr. HN Sethna took over the development efforts and conducted the first nuclear test in 1974. Dr. Raja Ramanna, who was also instrumental for the nuclear test became the Chairman of Atomic Energy Commission and gave a strategic map for the country. **Dr. Bhabha is the reason behind India being at an advanced stage of research in nuclear energy today.**

Pakistan invaded Kashmir in what they call their operation as 'Operation Gibraltar' in the year 1965. It led to a full blown war, and Pakistan was finally defeated. By this time Pakistan had started to get substantial Chinese material and diplomatic supports. This relation got forged with a basic "Chanakya's" concept of "adversary's adversary is a friend". Pakistan was also showing great interest in developing a nuclear weapon by this time.

In the 1971 war between Pakistan and India, China again supported Pakistan. Before entering the erstwhile East Pakistan (Present Bangladesh), Indian forces were required to substantially divert their resources to check the Chinese movements in the border. When Pakistan was comprehensively defeated and its territory dismembered, Pakistani-Chinese relationship had matured well enough to be called as a military alliance. Mr. Bhutto who became the leader of Pakistan in the year 1972, earlier in the year 1965 itself had proclaimed that by all means Pakistan had to go nuclear. He is reported to have said "Pakistanis will eat grass but make a nuclear bomb".

After the disastrous defeat of Pakistan, China is said to have aided Pakistan greatly and rendered support to subsequent actions required to go nuclear. The situation became critical. Because on one front China is a nuclear and missile power. On the other front, Pakistan overtly wanted to become a nuclear weapon state. To protect from the history of foreign invasions repeating on India due to India's military weakness of the past, India was forced to step in to armament programmes.

So the "GO" command for developing a nuclear weapon was given by the Indian Government. At the time when authorization was given, the technology for developing a nuclear bomb largely existed. The technology gaps were plugged in. India conducted the first nuclear weapon test in the year 1974. Further tests in the year 1998 made India a self declared nuclear weapon state. Both tests were possible only because of the strong leadership and courage exhibited by the then Prime Ministers Mrs. Indira Gandhi and Mr. Atal Bihari Vajpayee respectively. Yet the world must know India is the most responsible nation. India has never parted nuclear technology with any nation nor sold nuclear materials. The Indian nuclear policy is clearly defined.

India's Nuclear Doctrines

India's nuclear doctrines are No First Use, Non Usage on Non-Nuclear States and Essential Minimal Deterrence.

1. ***No First Use:*** In the event of war against a nuclear nation India will not bother one to use a nuclear weapon first. But if attacked with nuclear weapon, the enemy will receive a punishing strike beyond the level of imagination.
2. ***Non Usage on Non-Nuclear States:*** India will never use a nuclear weapon on a non-nuclear state - unlike USA's nuclear attacks on Japan killing civilians in thousands just in a flash.
3. ***Essential Minimal Deterrence:*** The amount of nuclear stockpile maintained will be just sufficient enough to maintain a minimum deterrence.

These ideologies show India's courage and compassion towards mankind even including the enemy! Such a nuclear doctrine, only India can formulate because of the rich heritage followed through the millennia of a non-aggressive civilization.

Need for Delivery Systems

After developing the nuclear weapon in the year 1974, it was felt by the concerned people that in the event of a nuclear strike on India, the primary requirement would be that the retaliatory forces should first survive the enemy's nuclear strike. As for retaliatory nuclear strikes, first their survival is essential. So our forces should possess a dependable and survivable nuclear delivery means that would survive the enemy strike.

All over the world, during the fifties and sixties nuclear weapons were largely to be dropped only by aircraft. But the seventies saw an enhanced development of air defence systems against aircraft. Hence, to drop a nuclear bomb, the aircraft had to penetrate the enemy air defence, go over the target and if at all it survives on its flight, drop the bomb. So a viable and dependable means of nuclear strike delivery was required and that was ultimately the ballistic missile. In the seventies, ballistic missiles had replaced long range bombers and had been widely fielded by major powers.

So in our case to return a punishing nuclear strike, ballistic missiles were required. But no nation would sell us such weapons after proving our nuclear capability, even though we required ballistic missiles.

So, in pursuit of missiles, an indigenous, accelerated ballistic missile development program was embarked on.

3

IN PURSUIT OF MISSILES

Our continuous efforts to develop different types of missiles and their technologies paid off. It was no small effort. It was a heroic effort to develop lethal missiles at the most difficult times of our nation, without any technological assistance from other countries.

THE EARLY EFFORTS

Defence Research and Development Organisation, well known as DRDO was formed in 1958 with Dr. DS Kothari as the Chief. The objective of DRDO was to promote defence research in support of development of weapon systems and equipment for the Indian Armed Forces. As missile technology was a growing avenue, DRDO embarked on missile projects and that too long before India going for nuclear weapons.

In those early days itself, DRDO formed a Guided Missile Study Team which was later renamed as Special Weapons Development Team (SWDT). This SWDT was to study the possibilities of guided missile development in India. Initially SWDT was located in New Delhi and later moved to Hyderabad in June 1962 as Defence Research and Development Laboratory (DRDL). In the later years, DRDL became the premier establishment for missile development.

SWDT came out with a feasibility study to develop a first generation anti tank missile and the go ahead was given. Developmental works started for developing an Anti-Tank Missile (ATM) in the year 1964. The efforts picked up momentum in the year 1966 after the India-Pakistan War in 1965. During the next five years time, a first generation Anti-Tank Missile got developed. These missiles while flying would unspool a wire for its entire range of flight. An operator will direct the missile's flight through his console (something like a remote control). These steering commands will reach the missile's guidance system through that unspooled wire. But the harder part is to track where his target is moving, also track where the missile is flying and guide the missile to hit the target. Such was the technology of those days; it was a great leap for an impoverished nation. The missile went through several successful flights. Unfortunately the Indian Army was not interested in this missile since second generation anti-tank missiles were available in the international arms market. So this ATM project was closed.

During the course of time many personnel came in and went out of DRDL. A study was initiated on the design of multi stage Surface to Air missile. With the strong support from Indian Air Force, in the year 1972, DRDL started a project called 'Devil'. It was a Surface to Air Missile (SAM) based on Soviet Union's SA-75 SAM system. The SA-75 was then being operated by the IAF. The development process of 'Devil' was based on reverse engineering. A fresh missile would be disassembled and a particular section would be taken out. A replica of the section taken out would be developed by copying up to the degree of nuts and bolts. Then this replica would be integrated with the original missile. The missile would be fired and if it successfully flies, then the replica developed was considered working and would be considered a success. In the due course of time, wherever possible some contemporary components were used. Slowly the entire missile components were developed. A stage was reached where an entirely new missile developed by DRDL was ready.

Around this time, parallely research projects were carried out to develop Inertial Navigation System (INS), High Process Missile Components, development of a 3 ton and 20 ton bi-propellant liquid engines. These engine developmental works fetched fruits during the Prithvi missile development.

When 'Devil' flight trials achieved the success, the IAF started phasing out the SA-75 system itself. So the project was closed in 1980 driving the DRDL scientists to low spirits.

By this time, DRDL initiated competence building projects. These projects focussed on missile technology developments like Ramjet propulsion, Target Seeker etc. But unfortunately, the closure of ATM and Devil projects led to the demoralisation of the pool of missile scientists. On its part, the Public Accounts Committee

(PAC) reviewed DRDL, since organization was producing no results.

During this intervening period, Dr. Raja Ramanna from the Department of Atomic Energy, was made the Scientific Advisor (SA) to the Defence Minister in 1978. He sensed the successes of Indian Space Research Organisation (ISRO) and Department of Atomic Energy (DAE) were not complemented by DRDO. ISRO had been regularly succeeding in its endeavours. Lately it had achieved launching Rohini satellite through the indigenously developed SLV-3. DAE had detonated the nuclear weapon successfully under his leadership and was galloping in many research areas in nuclear technology. Dr. Ramanna knew that the DRDL was not short of talent or competency but only in short of leadership. So he decided that he could bring Dr. Kalam, the then Project Director of SLV-3 to the missile development projects to infuse new culture and leadership to fetch results. This was a great move by Dr. Raja Ramanna to put a strong foundation for the missile programme.

I was working in ISRO with Dr. Kalam all through the successful days of SLV-3 and his troubled days at 'Aerospace Design and Dynamics Group,' Vikram Sarabhai Space Centre (VSSC). VSSC formed a study group for Polar Satellite Launch Vehicle (PSLV) and a configuration was evolved. The original proposed configuration by VSSC was a huge vehicle weighing 400 tonnes with two large strap-on boosters. Prof. Satish Dhawan realised that the proposed configuration was complex. He wanted to develop a reliable PSLV which would be the workhorse of ISRO and to be the core for the future launch vehicle projects in the years to come. So Prof. Satish Dhawan instructed Dr. Kalam to evolve an alternate configuration for PSLV with a team to support. Dr. Kalam's team with me and some other colleagues worked on an alternate configuration. After two years of continuous work under the guidance of Prof. Satish Dhawan, we evolved a configuration which was quite different and much lighter from what had been originally suggested. Today I am happy that the configuration evolved from our team's tireless efforts, has become the main stay for ISRO to have successive successful launches injecting satellites in the polar orbits.

In Dr. Kalam's own words in his 'Wings of Fire',

"I recall my working at ISRO HQ, Bangalore, as Director, launch Vehicle Programmes/System, in the early 1980s, when we were debating the performance and cost-effectiveness of launch vehicles. In 1981, the scientists of VSSC, Thiruvananthapuram, with the help of other ISRO centres, evolved a configuration of the PSLV core vehicle with two large strap-on boosters. The PSLV weighed about 400 tonnes at take-off. Prof. Dhawan wanted to study an alternative and simple configuration. I and some of my colleagues, A.

Sivathanu Pillai, N. Sundararajan and K. Padmanabha Menon, carried out mission, technology and feasibility studies for the optimal configuration. The team designed several options, including a unique core vehicle with an advanced solid propellant booster, using first stage rockets of SLV-3 as strap-ons. This brought the PSLV weight down to only about 275 tonnes at take-off."

We were also thinking of a missile project called REX (Re-Entry Experiment). Dr. Kalam wanted to convert the SLV-3 into a missile. But Prof. Dhawan wanted to keep ISRO away from missile development as it might affect the international cooperation that ISRO had with many countries. But Dr. Kalam's mind was always revolving around developing a re-entry class of missiles. Indeed he did that, not in ISRO but in DRDO. Do you know what missile that was? It is none other than our "AGNI".

In Dr. Kalam's own words regarding REX in his 'Wings of Fire':

"Meanwhile, I carried out an analysis of the application of SLV 3 and its variants with Sivathanu Pillai, and compared the existing launch vehicles of the world for missile applications. We established that the SLV 3 solid rocket systems would meet the national requirements of payload delivery vehicles for short and intermediate ranges (4000 km). We contended that the development of one additional solid booster of 1.8 m diameter with 36 tonnes of propellant along with SLV-3 subsystems would meet the ICBM requirement (above 5500 km for a 1000 kg payload). This proposal was, however, never considered. It nevertheless paved the way for the formulation of the Re-entry Experiment (REX) which, much later on, became Agni."

In ISRO, in spite of SLV-3 success, Dr. Kalam, was facing frictions - some of his bad times. He was thinking of moving to missile development, utilizing the offer of Dr. Raja Ramanna. But Prof. Satish Dhawan did not approve this. Dr. Kalam wrote a hand written letter to Prof. Dhawan expressing, "My heart is on missiles and I want to go." Dr. Raja Ramanna, passed his observations regarding Dr. Kalam to his successor Dr. V.S. Arunachalam, who approached the then Defence Minister Mr. R. Venkataraman and got Dr. Kalam to DRDL from ISRO. A new era in the Indian missile development began.

THE IGMDP

In DRDL, Dr. Kalam chaired the Missile Study Team (MST) to evolve the types of missiles required for Army, Navy and Air Force. His committee included members from R&D, Production and Services. The study team came out with five projects namely 'SS 150, SAM-X, Tactical core vehicle, 3rd generation anti-tank missile and a re-entry technology demonstrator', to be undertaken. The feasibility study in what is called as Guided Missile Development Program (GMDP) was presented to the then Scientific Advisor to Defence Minister Dr. VS Arunachalam, the then Chiefs of Armed Forces and to the then Defence Minister Mr. R. Venkataraman. The defence minister instructed the missiles to be developed concurrently instead of sequentially. Hence, the project's name was changed to "Integrated Guided Missile Development Programme (IGMDP)". IGMDP is the great success saga of India.

The Integrated Guided Missile Development Programme (IGMDP) with five Projects was formally sanctioned by the Government in July 1983 with duration of 12 years. They are:-

1. SS-150: A tactical surface-to surface missile of 150 km range for the Army, later named 'Prithvi'.
2. Multirole-tactical core: A multi-role missile system for the Army, the Navy and the Air Force, later named 'Trishul'.
3. SAM-X: Surface to Air missile (25 km) with multiple target handling capability for the Army and the Air Force, later named Akash.
4. ATM-3: An advanced third generation anti-tank missile for the Army, later named Nag
5. RTV: A re-entry test vehicle to develop re-entry technology, later named Agni. The technology demonstrator was later developed into a full-fledged weapon system.

The task of developing five missiles simultaneously, within short span of time was not so easy. It required a great deal of reorganization of DRDL for an effective management structure. During this critical period, Dr. Kalam used to frequently visit Bangalore almost on every Sunday. He did this to meet his ISRO friends in Bangalore and especially to meet me to discuss the various techno-managerial issues of IGMDP along with a lunch at the 'Woodland Hotel'. In due course he made me to be a part of his decision making process and wished that I could join him in the missile group.

Obviously Prof. Dhawan, who had been grooming me, did not agree in letting me shifting to DRDL. He also rejected the suggestions from Dr. Kalam many times and also informed me. But I knew that Dr. Kalam wanted me desperately to strengthen him at DRDL. This was because he had doubts of achieving results in DRDO with the culture existing then. For Dr. Kalam, having worked with the wonderful ISRO culture, it was extremely difficult for him to transform a regimental cultured DRDL, to achieve expected results. Many times, he expressed his worries. He needed the help of ISRO in many ways to support him in terms of technology reviews by experts, solid booster of SLV-3 for Agni, use of Sriharikota range for launching missiles and so on. Fortunately, my availability at ISRO HQ enabled him to get all the required support.

After Prof. Dhawan, Prof. U R. Rao became the Chairman of ISRO. Now both Dr. Kalam and Dr. Arunachalam were successful in getting me to DRDL on deputation. When I arrived at DRDL, the design process was just over for the Prithvi and Agni missiles. But yet lot of developmental tasks were required to be done to accelerate the projects.

I became the Director of Planning and Programme Analysis. The purpose was to give a new accelerated thrust for technology development, networking different groups as well as programme management for the five projects. I was required to help Dr. Kalam to restructure DRDL into a unique matrix set up. In this new structure the importance of project directors was increased and they were empowered with more administrative and financial powers, as it was in ISRO. In DRDL the programme management culture did not exist. My ISRO experience and working closely with Dr. Kalam came in handy to bring an effective programme management system that could help to streamline critical technology development process. In a multi-project environment with limited resources and conflicting interests, programme management was very much essential to achieve the objectives of IGMDP.

Objectives of IGMDP

Dr. Kalam framed the objectives for the IGMDP with a success mantra. These objectives will apply for any technology development program. The objectives were that "Once the missile is developed, it should have contemporary performance at the time of deployment with the following features:

1. 'Futuristic' - in terms of concepts and technology
2. 'Multi role' – different capabilities
3. 'Multi user'– to be used by Army, Air Force and Navy"

Futuristic

Dr. Kalam was very sure that the user would reject the missiles being developed in the IGMDP, if they were not contemporary when compared to the performance of missiles available in the world at that time. So a missile if it has to be contemporary at the time of deployment, it should be futuristic. The sad history of the 'Devil' and 'ATM' projects was not to be repeated. In the Devil and ATM missile development the technology was already obsolescent when the development started. When the development ended and a weapon system matured they had ultimately become obsolete. 'Military technology' advances very fast such that the state-of-the-art technology becomes obsolete in a few years. Moreover, the systems available with the armed forces were already obsolescent because no nation provides a weapon system of the same class and quality used by its own armed forces. Only export quality systems with inferior performance are offered to other countries. What is export quality?

In commercial commodities export quality is the most superior quality possible for that commodity. But in the arms market, export quality means degraded version of the original weapon system. Secondly, huge amount is charged for those systems which are highly demanded in the world market.

So for IGMDP, keeping these things in mind and going by the proverb 'Successful men work today for tomorrow', we started working for futuristic missile systems. By working for a futuristic system, the Indian Armed Forces would get a superior and contemporary system at the time of deployment of that weapon system. It means that there was a need to leap frog in futuristic technologies, deviating from the beaten path for which new concepts, design methodologies and management approach had to come together.

Multi-role and Multi-user

A multi-role and multi-user system reduces the need for different systems for various missions and for different users. A single system can be used for various missions. The benefits resulting from a multi-role system are greater.

From the angle of production, this in turn benefits both the producer and the user. For the producer, he gets more jobs and also more profit. For the user, he gets the advantage of cost reduction due to the mass production.

The user can maintain a large force level and need not use different systems for different missions.

For user, the training and maintenance becomes simpler. Instead of maintaining different types of systems in his inventory, he can just maintain a single type. It greatly reduces the menace of maintenance. Instead of training personnel to maintain diverse systems the user needs to train personnel to maintain only one system.

Such advantages do exist for the user due to a multi-role system. And we decided that all the five missiles must have multi-role capability. For example, the Trishul missile had three roles of being an anti-aircraft missile (ground to air), anti sea-skimmer (missile fired from ship against anti ship missiles) and as anti-radar missile (air to ground) and hence got the name meaning the three bladed 'Trident'. Similarly, Prithvi has field interchangeable warheads, where the users can choose warheads and fix them in the field according to the type of targets to be neutralized. Prithvi can be used from mobile launch system on ground or from ship against land targets and used by all the three services. Since Prithvi is a surface-to-surface missile, it was named Prithvi. Similarly, medium range surface to air Missile Akash is on tracked vehicles and wheeled vehicles suitable for use by Army and Air Force respectively. Akash can be used as SAM against Aircraft and was also to be used against incoming ballistic missile like Patriot. It was named Akash as the theatre of war is in sky (Akash) with multi target engagement capability. Nag is a third generation anti tank missile making a top attack on the enemy tank like Cobra. It can be launched from a tank or from helicopter. The long range Agni has to go to space and re-enter in the atmosphere. At the re-entry point, the missile module experiences a very high temperature more than 2500°C which is equivalent to Agni Pravesh and hence it is named Agni.

Was it possible to realize these objectives? Was it a cakewalk?

CONSTRAINTS AND DIFFICULTIES

Yes, it was possible to realize the objectives but definitely was not a cakewalk. There were difficulties and many harsh constraints.

Difficulty

The biggest difficulty was the existence of a huge technology-time gap of more than 30 years in these areas as compared to the developed Nations who have a technological advantage even before World War I. During the cold war between Soviet Union and USA, the technological progress was tremendous. Though we are proud to say that we have used the world's first war rocket in 1792 in Srirangapatna War, but in the 1980s when we were starting the IGMDP, the technological gap was enormous. The first world Nations had the advantage of this time gap, which we wanted to nullify through IGMDP. It is a tall order in any sense. But we were sure, that we could do, provided we were able to forecast newer technologies not in the line of thinking of other nations.

We also realized that industrial development in the country at that point of time was not conducive for R&D to production. The colonial rule kept India away from the industrial revolution. This had serious implications of non-availability of suitable industries in India for taking up major missile development programmes. Post 1962 Indo-China war was the licence production era. Only the necessary knowledge to manufacture would be infused by the producer nation to Defence Public Sector Units. The disadvantage of the License Production is that the Industry will have only 'Know-How' capability and not the 'Know-Why' capability. In real sense, Licence Production obtained by the defence PSUs had exposed to only very little technology. The only silver line India had was the thrust given by Prof. Satish Dhawan for the industries to participate in the space programme. ISRO's contribution to create a minimal base in the country for aerospace technology products were to be exploited for IGMDP.

Constraints

The constraining factor prevalent for India to break the technology gap was the 'technology denial regime' enforced by first world nations. This regime is called as Missile Technology Control Regime (MTCR).

Missile Technology Control Regime: What is that?

Missile Technology Control Regime (MTCR) is an informal and voluntary association of various countries to prevent the proliferation of unmanned delivery systems capable of carrying a 500 kg payload (read nuclear warheads) to 300 km. MTCR seeks to coordinate restrictions in exports like complete rocket systems, unmanned air vehicles, and related technology. Unfortunately some of the systems and technologies having greater civilian applications are denied or come with a strong end user agreement that it would not be used for any military purpose. It was originally established on 16th April 1987 by Canada, France, Germany, Italy, Japan, UK and USA. In the year 1988 India test fired Prithvi and the formation of the regime in the year 1987 cannot be any coincidence. Russian Federation became the signatory in the year 1995 and China has verbally pledged that it would adhere to the MTCR in the year 1991.

The MTCR was initiated in response to the increasing threat of proliferation of 'Weapons of Mass Destruction' (WMD), by certain countries (read as India). So Nations that wish to develop missiles to protect themselves are controlled by MTCR, by denying technologies and systems to them. MTCR has a huge list of technologies and systems that are banned from exporting. The regime leaves the signatory nations to restrict them from exporting these listed items.

Some of the common systems like the Computerized Tomographic scanning systems (commonly called as CT scans) come with an end user certificate. If the system is found to be used for military purpose then that organization will be blacklisted and will be denied further imports. Many of the military grade computing processors are also restricted by this regime. Military grade processors are highly reliable and can withstand a variety of physical constraints like extreme temperature, etc. These military grade processors can find applications in civilian projects but are yet restricted. But all such transfers are considered on a case by case basis (read as nations that are supported by big nations can import these technologies and proliferate missiles).

India is not a signatory of MTCR and many other controversial treaties like Non-Proliferation Treaty (NPT) and Comprehensive Test Ban Treaty (CTBT). But India is a signatory of the Chemical Weapon Convention (CWC) and Biological Weapons Convention (BWC) that demand the ban of Chemical and Biological weapons. India is completely against chemical and biological warfare and has behaved responsible in every world forum.

But India is determined neither to proliferate nor to support proliferation of these weapons. The goal is to make the world free from weapons of mass destruction. But those days are very far or may not even come. But India has a clear cut NBC Policies that have been recognized worldwide as the most responsible.

Through MTCR, many technologies that could have been easily procured from abroad were denied to us. When some technology or equipment is available, then it is obvious to use it as it saves time. On the other hand, developing the same will be time consuming and foolish; this is called as "Reinventing the Wheel". But unfortunately we were forced to such a situation to reinvent the wheel; in fact we reinvented so many wheels!

Also when the economy of the country was not flourishing, pooling money in such issues was a huge constraint. But the economical boom after the early 1990s helped to overcome many such difficulties. At this point I tell you, "Unfortunately we missed industrial revolution due to foreign occupation". Had we taken part in the industrial revolution the international stature of India would have been unimaginable. "Fortunately we have taken part in Information

Revolution and it has nourished us. Thanks for the time being on our side."

In the hindsight MTCR will look like a curse imposed on India. Is it so?

Result of MTCR

There is a saying "Blessings in Disguise". MTCR was definitely a curse, but there was hidden bliss in that curse. That bliss is nothing but technology empowerment. We developed all the technologies that were denied to us. How? The nations which imposed the technology denial regime saw DRDO just as a single small entity. They observed DRDO as an organisation meant for developing missiles. But they missed to see that there existed a bigger entity, the "INDIA". India is the land from where the 'Panchatantra' emerged. We decided to follow one of the principles of the Panchatantra – Unity. Our nation stands on the foundation 'Unity in Diversity'.

The presence of a strong intellectual and scientific community was the National Strength. The required expertise was available but distributed in parts in many research labs, industries and greatly in Academic institutions. When we networked all these entities - the industry, laboratories and academia with IGMDP, we got the necessary expertise. The existent two decade experience in rocket engineering led the way. This certainly proved us "Unity is Strength". With this strength it was possible for technology empowerment, defeating the MTCR and we proved the world that "TECHNOLOGY DENIED IS TECHNOLOGY GAINED".

TECHNOLOGY DENIED IS TECHNOLOGY GAINED

Resultant of the MTCR, all the systems had to be home grown. Achieving technology independence through IGMDP was like a mini freedom struggle and at last the national strength won the technology denial regimes.

Nearly 200 technology packages were identified to be developed for the success of IGMDP. All these packages were developed for the first time in India. The consequence of any failure in developing them would directly result in the failure of IGMDP itself. Building critical technologies were full of challenges, because they were new and the base knowledge was insufficient. The knowledge, manpower, infrastructure and industrial support had to be made available in the required manner according to the technology being developed and its relevant importance in the IGMDP. Then the progress of the development had to be constantly reviewed and problems had to be solved. Technology empowerment became possible through a consortium approach and also through the collaborative partnership among the identified partners from R&D labs, academic institutions, and industries. Those activities had to be planned and strategically managed in order to realize the critical technologies.

Above all these measures, we had to be innovative. When we went out of the beaten tracks, success came like a fruit dropping into our laps. I remind you the objectives of the IGMDP especially the missiles should be futuristic. For the missiles to be futuristic, the technologies should be futuristic. But the technology we had at that time was only rudimentary. So we had to develop the required technology only with the resources available at that time. How to reach the 'futuristic' objective? We simply have to be innovative. By being innovative we developed the technology packages to the required standards.

LEAP-FROGGING THROUGH INNOVATIONS

Though we were able to develop the missiles we still had to adhere to the most important objective of being futuristic. We knew that the missiles that we would develop would be of state-of-art at the time of deployment. We substituted innovation against superior technology. This innovation carried certain risk elements but if we succeeded then the missiles would become futuristic. Emboldened by our objective of making our missiles futuristic we decided to take the risk that imposed a 50-50 chance of success. The risk taken by us paid off.

I am just narrating only two important technology developments in India combating the MTCR. These include Re-entry heat shield development for Agni, and Guidance system development for Prithvi. These technology developments ultimately raised our technological capabilities to international standards. This elevation to international standards paved the way for Joint Ventures, where the technological parity is required between the partners. IGMDP helped in such a fashion for the formation of the BrahMos Joint Venture.

DEVELOPMENT OF RE-ENTRY HEAT SHIELD FOR AGNI

When a ballistic missile reaches the edge of the atmosphere say an altitude of 100 km, the warhead gets separated and starts descending towards the target. While falling, it re-enters the atmosphere at a speed greater than 10 mach (1 mach = 330 m/s). The warhead experiences a stiff air resistance and due to the resulting friction, it gets heated up to a mammoth 2500°C. When we speed up even in ordinary buses and motorbikes we can feel the air resistance in headwinds. Imagine the friction at a speed greater than 10 times the speed of sound.

The warhead contains explosives which is conventional or nuclear material. If such heat is allowed to seep inside, the warhead would go useless. So, thermal insulators are required to protect the warhead from heating. I hope one will never forget what happened to Kalpana Chawla. She and her crew were returning to the earth in the space shuttle 'Columbia'. At the time of re-entry, the failure of thermal insulators resulted in the loss of the shuttle and the crew. Agni also has to endure this re-entry phase.

Now how to develop such thermal insulators? Metals generally melt and evaporate at these high temperatures. But such insulations are possible through carbon based composite materials for the structure and carbon-carbon composite material for the tip. This material has the ability to ablate at high temperatures and high velocity in a uniform manner. Due to this the flight trajectory is maintained, protecting the warhead by maintaining the temperature at less than 50 deg C. This technology is essential for any long range missions like Inter Continental Ballistic Missiles (ICBM) or any manned space missions. Re-entry technology is therefore a highly guarded secret by the few nations possessing it. A team of scientists headed by Mrs. Rohini Devi took it up as a challenge and successfully developed the re-entry technology.

While developing any flying machine how can we know about the aerodynamic behaviour of it? One option is to test the scaled down models in wind tunnels. In these tunnels the model is kept stationary and air is blown on it at the speed which that vehicle is expected to fly. The flying conditions are simulated at various speeds at which the vehicle is expected to fly. In this wind tunnel, various aerodynamic performance parameters like the drag at particular spots in the body are studied to validate the design.

Now recall the trajectory of a ballistic missile; the vehicle re-enters the earth's atmosphere at hypersonic speed. To validate the aerodynamics of the nose tip, we needed a Hypersonic Wind Tunnel - one that blows air at hypersonic speeds. We did not have such a facility.

The other option was to arrive at the optimum shape through Computational Fluid Dynamics (CFD), which helps to validate the aerodynamic design of a flying machine. The aerodynamic design comprises of thousands of grid points; assume these grid points as a node in the skeletal structure. CFD software validates the design by calculating the mathematical parameters of each grid point. But before the advent of computers, they were all done through mathematical calculations. Manual mathematical calculations for nose tip evaluation would require years to complete. In the 1980s we had just stepped into the information age and mathematical calculations were done in desktop PCs, which required at least 9 days for evaluating and optimising one grid point in the nose tip among thousands of grid point. So we required a supercomputer. But, from where do we get a super computer? US made super computer 'CRAY XMP' was available at 'Indian Meteorological Department' which was used for weather prediction.

And now came the hindrance; the computer was procured from USA with a restriction that it should not be used for any other (read as defence) purpose. So in spite of the availability of the machine, we were unable to use it. Dr. VS Arunachalam, the then chief of the organization and Dr. Abdul Kalam, the then chief of missile programme went to USA and made a request to US Defence Secretary Mr. William Parry for the use of the existing or agreeing to sell a new super computer for defence R&D purposes. The Defence Secretary gave a good lunch to them but not the super computer. Returning with empty hand, Dr. Arunachalam established Advanced Numerical Research and Analysis Group (ANURAG) with Prof. G Venkataraman deputed from IGCAR, Kalpakkam as the leader to design and develop supercomputer for undertaking the Agni re-entry module design, in 24 months.

Right, created ANURAG but how to achieve a supercomputer? Till date we are buying ordinary computers from abroad and not manufacturing any microprocessor at home. Then imagine the scenario some 25 years before. That was the age when ordinary calculators were treated like computers among the general public of India. When there was no indigenous capability to develop an ordinary computer how could we develop a supercomputer? You will be astonished to know that Dr. Venkataraman, Dr. Neelakantan and fourteen youngsters developed a Hypercube based parallel processing supercomputer for the CFD analysis. In the meantime another group was established at Indian Institute of Science (IISc) headed by Prof. Deshpande to develop new software codes for the CFD.

Even with industrial computing systems and a new software code called BHEEMA, the calculation time was greatly reduced. While this progress was on, simultaneously hardware and software developmental efforts were on full swing.

Finally, in 24 months time, PACE+, a 32 node parallel computer got developed with a capacity of 1.7 Giga flop. Roughly 1 flop can be compared to one calculation. PACE+ was capable of executing 1700000000 floating point operations per second. It was 20 times faster than the 'CRAY XMP' which was refused to us. With the combination of PACE+ and the new software codes for CFD, each iteration got reduced to just five minutes from 9 days.

The nose tip model was evaluated and cleared for integration with the missile. The irony was that after the development of our super computer, USA offered their CRAY XMP which we promptly declined. Today we have mastered the re-entry technology and Agni has got various versions having ranges from 700 km to 5000 km. Thus, innovatively developing the supercomputer and the usage of home grown software paved the way for success.

DEVELOPMENT OF INERTIAL GUIDANCE SYSTEM FOR PRITHVI

Prithvi is a very uniquely designed short range ballistic missile. It is not that we were the first to develop a short range ballistic missile. Many countries have developed such missiles much ahead of us. But what makes Prithvi unique? It's manoeuvrable trajectory.

Prithvi has to manoeuvre and also be highly accurate to reach the target which is decided by the accuracy of the guidance system. Such guidance systems are entirely dependent on the high-precision navigational sensors. As usual, this was not available to India. Without these sensors there was no other possibility for developing a high accuracy guidance system. Four young scientists were given the tasks to come out with the solution, which would enable achieving better accuracy using the rudimentary gyroscopes that were being produced in Hindustan Aeronautics Limited (HAL).

This led to a great idea of error compensation. Those scientists studied the sensors and modelled that how the error in sensing occurs. They did many mission runs to characterize the sensor and calculated the errors at various stages of the flight. The simulation of the trajectories including many perturbations and failure modes was carried out to estimate the error bands. The idea was to compensate the error by making the guidance computer understanding the error. When the computer understands the error, it compensates for the error. How to compensate was instructed through the software so as to give better accuracy at the end of the flight. The system had got validated in hardware-in-loop simulation and later in the flight trials. The innovation of compensating the hardware induced error through software played a key role in realising an excellent inertial guidance system.

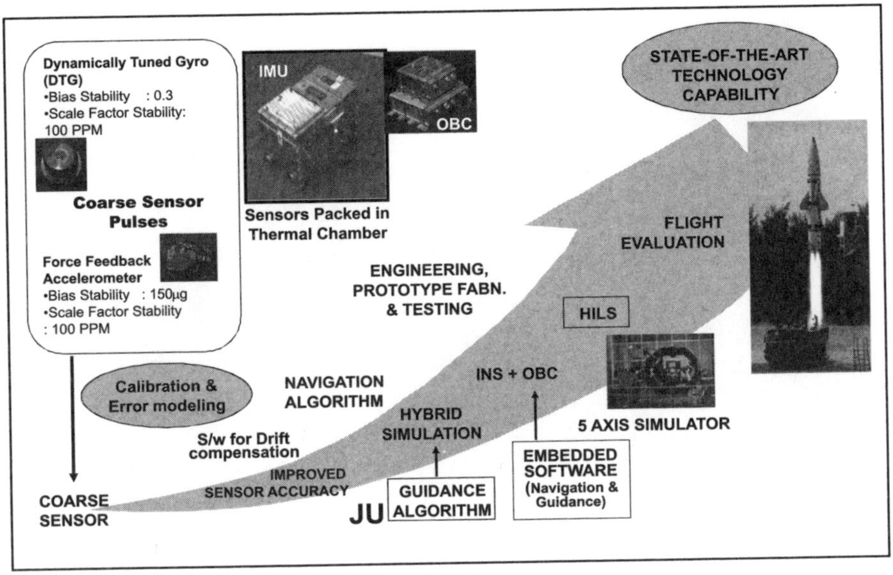

THE END RESULT OF IGMDP

The result of IGMDP made India a nation with self-sufficiency in strategic missiles, competence to design and develop any type of tactical missiles and most importantly the availability of critical technologies, missile scientists and infrastructure. Many off-shoot missiles came out of IGMDP. The value addition was the availability of innovative young minds to think big for the future. Many spin-off products came out of missile technology.

PRITHVI

Prithvi is a short-range surface-to-surface missile having high accuracy and manoeuvrable trajectory. What is Maneuverable Trajectory?

So far, the nations have designed their ballistic missiles to fly in the conventional parabolic trajectory. When a ballistic missile is inbound for an attack, the enemy missile defence system tracks it continuously and predicts the missile's trajectory. Then it calculates a suitable interception point. It launches its anti-ballistic missile (interceptor missile) such that at this point interceptor missile either collide or blast near the incoming missile to destroy it. For all these activities, enemy has a very very short time window. Failing on which, the incoming missile will finish its job. For the interception to be perfect, the missile defence system's accurate tracking is very important.

For the IGMDP objective, 'futuristic' Prithvi should be able to evade

Prithvi P-I
(Army Version)

Prithvi P-II
(Air Force Version)

Dhanush
(Naval Version)

anti-ballistic missile missiles. For that, Prithvi has to manoeuvre. A cruise missile can manoeuvre but for a ballistic missile, manoeuvring will be very difficult. So we thought of giving wings to Prithvi. The only ballistic missile in the world that has wings is Prithvi. With the wings, Prithvi can manoeuvre in the 'Yaw and Pitch' planes.

Because of this manoeuvre, the enemy missile defence system will not be able to predict the trajectory of the incoming Prithvi missile and determine Prithvi's targets it may manoeuvre at any time. This manoeuvre will confuse the enemy enough, so that he will always be at suspense that where the missile is going to impact.

So our implementation of manoeuvrable trajectory for a short range ballistic missile put India ahead of other Nations when those nations were deploying their first generation missile defence systems.

The ship launched a variant of Prithvi is called as Dhanush to attack any land target. The role of Prithvi and Dhanush was tactical / strategic missiles.

AGNI

In the IGMDP, Agni project was taken as 'technology demonstrator', to prove the re-entry control guidance. This was important as it was difficult to develop the long range missile in the first step itself. For

'Agni Technology Demonstrator (Agni-TD)', Prithvi's propulsion system served as the second stage and the SLV-3's first stage from ISRO as the first stage. It was first tested at the Interim Test Range in Chandipur in 1989.

The nose tip of Agni is made of Carbon-Carbon composite materials which can withstand temperatures up to 2500°C. Advanced composite airframes of the missile are made of carbon fibre / epoxy and carbon fibre /phenolic resin systems. These airframes work on the principle of ablation in which the material sacrifices itself at 'high temperature and velocity' and provides sufficient heat insulation to the inside electronic circuits and warhead.

After the Agni-TD demonstration flight under IGMDP, the second stage was modified to solid propulsion system and named as Agni II. This two stage, 2000 km range Agni-II is fully operational. In the meantime, realizing the importance of a shorter range solid propulsion missile, the booster of Agni was used as a single stage for a 700 km range. This version is Agni-I and it has been operationalized. Agni-III is designed for a range capability of 3500 km with advanced guided system. After successful trials, Agni III has also been operationalized. Agni and its variants prove the mastery of re-entry technology for different range capabilities. Lately Agni-IV followed by Agni-V also has been tested. Agni is a rail/road mobile system capable to be moved anywhere in India and launched.

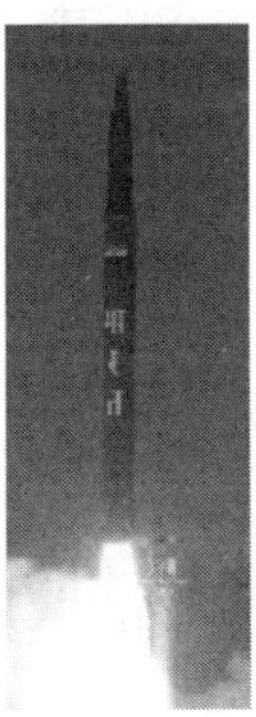

Agni I

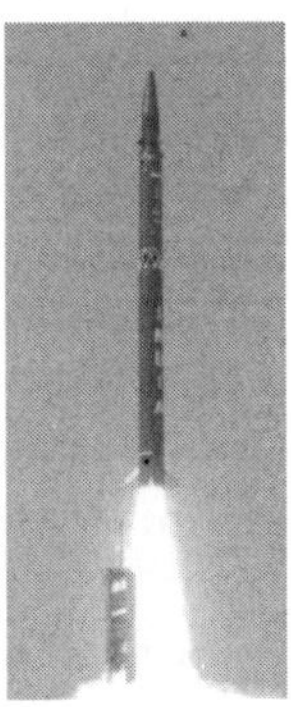

Agni II

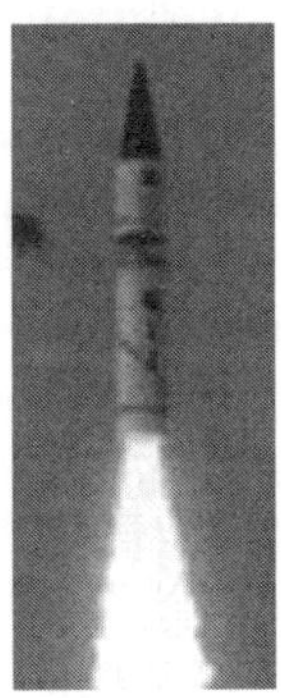

Agni III

Agni IV

Agni V

NAG

Nag is India's third-generation 'Fire-and-Forget' anti tank guided missile, with an imaging infra-red seeker. It can be launched during both day and night. The missile locks on to the target before being launched, and after its launch, homes on to the centre of the moving

target all the time. When it approaches the target, it finally lifts like a cobra and attacks the target. And hence it has the name 'Nag'. When Nag was under development, the contemporary technology at that time was wire guided, Second generation anti-tank missiles. The next generation is the wireless guidance to which Nag belongs. We infused an innovation to the missile, to make it more powerful. How?

A tank has very strong armour particularly to withstand anti-tank missile attacks. But the Achilles Heel of a tank is its top. The armour at the top of the tank is thinner than the other areas. So we made 'Nag' to attack the top of the tank to ensure complete destruction. Nag is being inducted in the armed forces. A helicopter launched version called as HELINA is also on the making.

TRISHUL

Trishul, the Sanskrit name for Trident, is a short range, quick reaction, all weather surface-to-air missile designed to counter a low-level attack. It has been flight tested in a sea-skimming role and also against moving targets. With a 5.5 kg High Explosive fragmented warhead, it has a range of 9 km. For target detection to missile launch it just takes around 6 seconds. The missile can engage targets like aircraft and helicopters, flying at speeds between 300 m/s and 500 m/s by using its radar command line of sight guidance scheme. It operates in the K-band (20-40 GHz), which makes it difficult to jam. In the K-band three-beam system, the missile is initially injected into a wide beam, which then hands it over to a medium beam, which passes over to a narrow beam, guiding it to the target. After a lot of flight trials with mixed results, the project was concluded as a technology demonstrator. Nevertheless the technologies developed for Trishul are being used in other projects.

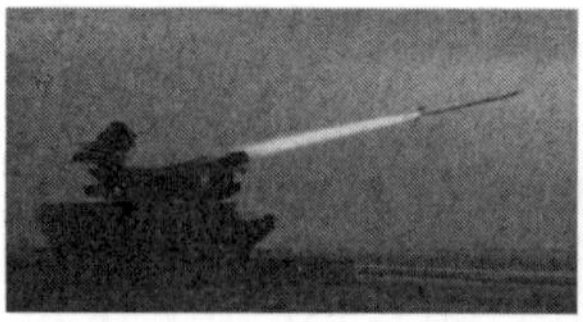

Army Configuration

Air Force Configuration

Navy Configuration

AKASH

Akash (Sanskrit name for Sky) is a medium-range, theatre defence, surface-to-air missile. It has got a multi-target handling capability and Electronics Counter Counter Measures (ECCM) features. It has got a range up to 25 km and can fly at an altitude of 30 m to 18 km. The missile is capable of multi-target engagement and with discriminating capability between friendly and enemy aircraft. This is a complete mobile system mounted on trucks in one version and on a tank chassis on the other version. Akash is guided by phased array radar which is based on phase shifters. The radar can track multiple targets and guide multiple missiles towards these targets simultaneously. Akash uses the Integral Rocket Ramjet propulsion and has a low reaction time of 15 seconds from detection to missile launch. It is capable of even detecting and destroying aircraft flying at tree-top height. Akash has gone through several successful flight trials and is accepted by the Indian Air Force for induction.

In anti-aircraft warfare, we expected that the trend would be linking all the missile launchers under a common command and control centre. Akash is a highly network centric system which is the trend for future war.

PROGRAM MANAGEMENT: THE KEY FOR THE SUCCESS

IGMDP was a multi project missile development programme, developing five different types of missiles, simultaneously. But due to the unavailability of the required technologies, IGMDP became a vast technology development program intertwined with missile developments. This intertwining made it one of the most complicated technology development programmes in the world. The technologies were new and futuristic. The technology base with DRDL was very limited. So a large number of government labs, academic institutions, and private and government industries were used. The technologies were to be developed at multiple institutions and laboratories with networked organizations as well as to make them not to have open ended approach. These technologies were closely linked with the realization of missile development.

Uncertainties, complexities, different working cultures, various kind of attitudes, military environment, secrecy, egos, high average age group laboratories, and difficult procedures - all attacked IGMDP at the same time. The PSUs had the culture of undertaking production with elaborate training given by foreign industries. They had difficulties to accept IGMDP culture. Private industries wanted a commitment in the beginning for production so as to make profit.

For tackling the difficulties in development, DRDL was reorganized in to a matrix structure of management. In a nutshell, the evolved matrix structure enabled each missile project being given the required technology system with an identified systems manager. Also the technologies to be developed were given a systems manager. The systems manager used to oversee the technology development in his regime whereas the project director used to oversee the entire missile project. They all reported to the programme management board that oversaw all the five missile projects. Each individual was to report to two bosses: the project director and the systems manager; through this matrix structure effective control was established. For example the systems manager would be tapped to know what was happening in that particular systems development, say INS, as he would be overseeing the INS development for all the other missiles. Similarly, the project directors would be tapped to know what was happening in that particular missile project. Only this matrix structure based program management helped to resolve conflicts, assign priorities, avoid duplications and enable lucid control of the process to stick to the schedule adhering to the objectives of IGMDP.

Inherent complexities of technology development, gave a real challenge for effective management. The complexities were multi-faceted - Networking of establishments and related management, Time criticality and Technology criticality and so on. Above all the multiple technology developments were intertwined with multiple missile developments.

Networking of establishments and activities monitoring was a hectic process in IGMDP. Today, we have internet that brings the world in our finger tips. But in those days internet was just a concept in India. In those days, computers were primitive with monochromatic monitors displaying green fonts in a black background and being operated only in air conditioned rooms. Even a television was a rarity, a home with an antenna (dipole antenna) served as navigation references in residential areas. The younger generation would not have heard about these antennae lest to see. Today it is the era of DTH with DVD quality video and audio. In that era, we had to find various industries, research laboratories and academic institutions manually through reports or personal contacts. We had to understand their expertise and capabilities and to delegate works and take them along with us.

In the IGMDP, as a whole we were developing more than 200 critical technologies and many sub critical technologies and so many less critical but important technologies. Everything was time bounded and we had to manage 22,000 activities including technology research, technology development and product manufacturing. All these activities were carried out across the length and breadth of India. Remember the communication facilities existed in India. People who have born after 90s may not be aware of the India of 80s. Owning a telephone or a television was a luxury and those who were financially affluent were able to afford them. Cars - they were a rarity unlike today where a person can buy a car and park it in his garage within a day. But those days one had to wait indefinitely for buying a car.

The programme management directorate at DRDL monitored 22,000 activities comprising of research, development, production and testing that were being executed at various places across India. We had to receive reports on those 22,000 activities at a month end, analyze them, understand the progress and difficulties, and address them. While giving corrective measures on the deviation or issues, we had to issue the next tasks in those activities for the forthcoming month also.

All those activities were time critical. All the developmental activities for a particular missile must coincide at a predetermined time; some may take even more time, and efforts had to be put in to see if that time could be reduced, how it could be done, feasibility study, error corrections, etc.

The missile development was time bound; almost each subsystem required some sort of technology development for it. The development of the missile cannot be stopped for want of that particular technology dependent subsystem. Worst case was, there could be so many subsystems requiring so many technologies for a missile.

Each second wasted would reduce considerable percentage of the user's confidence on DRDO's capability to deliver the missile – Just not one missile but five missiles. Along with the development of required technologies, it is essential to manage and solve personal issues, morale issues, material requirements, bureaucratic hassles, etc. It required an overall control scheme of the programme. The evolution of that scheme itself is a great outcome of IGMDP.

I will highlight some of the techniques used to progress IGMDP, using a great lot of experience I gained in launch vehicle projects at ISRO. We introduced different strategies for critical technology development, monitoring of large activities through a computer aided scheme, critical activity monitoring scheme with clear cut milestones. Conflict resolution in multi project environment with PACE technique, structured review system, flight trial mission management, capability build up, infusing technology skill, capacity build up by large scale induction of scientists reducing the average age of the organisation, continuous renewal by re-organisations in 1986, 1989, and 1992. The above activities introduced a new culture and a movement to achieve results. The seeds laid 25 years before in DRDL, have resulted in many young scientists managing huge missile projects and as Directors of laboratories today, and many generations will follow.

In spite of so many evil designs from outside pressures we as a nation indeed succeeded in our mission of technology independence. But we in DRDO felt that somewhere we were lagging. We felt that we have left an important missile system to be developed that would plug the gap in the battlefield.

AN UNFINISHED JOB

In IGMDP we concentrated on strategic offensive missile systems and tactical defensive missiles. Though initially conceived as a tactical missile, due to various reasons Prithvi became a strategic missile. In a war a strategic missile would be the final weapon of choice. In the initial stages, a strategic missile will not be used as it will give a wrong message. Because an enemy may not know what the strategic missile is carrying. It may be carrying a non nuclear warhead. But in general, the enemy would assume that the incoming missile is carrying a nuclear warhead. So he will retort with his nuclear missile strike.

A strategic missile serves only one purpose: DETERRENCE. It would be the message for the enemy that if he fires his strategic missile then we will also fire ours to finish him off. With ICBMs in their arsenal USSR and USA never fired them during their Cold War games. So falling in that league if Prithvi is fired in the war with a non-nuclear warhead the enemy will only understand the missile as a nuclear attack. Eventually he will retaliate and also he will have the excuse that he was only retaliating. Not only that we will be held responsible for provoking him though we had only launched a non-nuclear missile on him.

On the other hand, other missiles of IGMDP were purely defensive except Nag which could engage only tanks. So a wide gap was observed that an indigenous tactical war winning missile was absent in our country. A missile system which has to be the 'first strike weapon system' was absent in our arsenal. Our missile team used to discuss this issue often and tried to find a solution. But still we were not able to determine that "WHAT IS THAT FIRST STRIKE WEAPON?"

4

PERSIAN GULF WAR: THE SHIFTING IDEOLOGY

Taking it as a mission to motivate students - the pillars of Future India, I used to give lectures at schools and colleges at different parts of India. Many questions come from the students showing their curiosity, but only one question repeats in each corner of the country, "Why India must develop a cruise missile, and that too a supersonic cruise missile? Won't the present missiles like Agni and Prithvi do the job for our Defence?" This chapter will give the answer for many of you who might be having this probing question in your minds. I am going to take you for a ride on a time machine to the WW-II era and tell about a new 'paradigm of war' that turned centuries old doctrines upside down.

QUICKLY WINNING WARS: WHY?

From the ancient days prolonged wars were not favoured. Lengthier wars will erode the economy of a nation. A few sentences from the book 'Freedom at Midnight', written by '*Dominique Lapierre and Larry Collins*' deserve to be quoted. This book speaks about the condition prevailed in England during the year 1947, and depicts the after effects in a country that had won the prolonged WW-II.

The streets were almost deserted. The passers-by hurrying down their pavements were grim, joyless creatures, threadbare in old uniforms of clothes barely holding together after eight years of make-do and mend. What few cars there were, darted about like fugitive phantoms guiltily consuming Britain's rare and rationed petrol. A special stench, the odour of post war London, permeated the streets. It was the rancid smell of charred ruins drifting up like an autumn mist from thousands of bombed out buildings.

And yet, that sad, joyless city was the capital of a conquering nation. Only seventeen months before Britain had emerged victorious from mankind's most terrible conflict.

[Page-1, Dominique Lapierre and Larry Collins: Freedom at Midnight, Vikas Publishing House Pvt Ltd].

From the above quote, we can conclude that even a country emerging victorious from a prolonged war will be torn by the expenses incurred for conducting the war. And just for this reason only Tsun Tsui stresses about a quick victory.

A prolonged war will increase the war expenditure only, to slow down the growth rate of a country's economy. This drives that nation to a corner where a crucial balance has to be maintained in catering the dire needs of the society and the military. Citizens may also get agitated due to the prolonged conduct of war. These are the predominant reasons for countries not waging war when they are not sure about the time of winning. This uncertainty and incapability makes a nation to fight a proxy war by encouraging terrorism in other countries. Forging evil nexus to encircle and diplomatic dog fights are the result of poor economical strength to fight a prolonged war. I give the cost endured by USA for Invasion of Iraq and Global War on Terror.

USA spearheaded invasion of Iraq during 2003, did cost 616 billion dollars for USA as war expenses. In 2008, I led an Indian delegation to USA for exploring a possibility for any collaboration with USA in defence research. But the answer from their side was that presently R&D funds have been drastically cut for any research, as the war expenses have gone up. So for even a superpower – a powerhouse of economy – effect on its economy due to war was that much.

US economy further deteriorates due to their Global Police like attitude and policy. In the past 12 years, USA had to spend 1.2 Trillion dollars to sustain its efforts in Afghanistan. A trillion is 1000000000000. 1.2 trillion dollars is 1200000000000 dollars. Equating a dollar as 50 Rupees, the value is ₹ 60000000000000. This value is as much 5 times greater than the total expenditure of India for the financial year 2012–2013. This much of spending is because of the sustenance of combat in the harsh conditions of Afghanistan, and it is definitely affecting the US economy.

As a rule of thumb, the cost of war climbs with a harsher terrain, tough enemy and tough objective (like avoiding civilian casualties, etc.).

So to ensure quick victories, all over the world, new technologies are developed to build new weapons. Though war is the result of the conflict of interests between the nations, ironically it is the testing ground for technologies. After every war, new technologies are developed and based on them new war strategies are evolved and new weapons are fielded. So in every war fought, within a span of five years we can see improved technologies and doctrines being fielded. Air Power was one such doctrine evolved and followed during the WW-II.

EVOLUTION OF AIR POWER

From the ancient days, armies and navies were considered the mightiest forces. But in the WW-II, Air Forces emerged as a powerful arm with a potential even to decimate an entire nation with nuclear bombs. The air forces proclaimed with their chins up, "No matter how mightier armies or navies are, they are just sitting ducks for us," "Treat a bomb laden enemy aircraft on the air as a messenger of death." The evolution of the air power paved the way to speedy winning of wars.

The history of the origins of airpower will be interesting. After the invention of aircraft they were used for reconnaissance; there were no weapons in those aircraft. If two pilots from the opposite sides had to pass each other, they used to wave hands or otherwise threw odd things on the other aircraft to disable it. One pilot while flying in his aircraft saw an enemy aircraft and tried to shoot it with his handgun. This triggered some minds and they started fitting big guns in aircraft to shoot down the enemy aircraft.

One pilot during a mission somewhere in Africa thought, "Bullets don't break a nation but bombs do!" So he dropped small grenades on the enemy ground troops. The effect and success of this action triggered the aeronautical engineers to develop aircraft that could carry huge bombs and drop them over enemy's head. Later proper aiming devices were added to drop them accurately on the targets. The shooting aircraft was called as 'fighter' and the bomb dropping aircraft was called as 'bomber' and both are enemies. Still today both these aircraft share their enmity. WW-II saw the emergence of powerful fighters and bombers.

The credit for using airpower effectively should go only to Germans. They developed a doctrine called as "Blitzkrieg" in which, German Air Force used to come from nowhere and attack the enemy on the ground. Immediately after the aerial attack, the German tanks would come into action and engage the remaining enemies to decimate them. Classic examples of air power's effectiveness were the annexation of Poland and initial phases of invasion of Soviet Union by Germany. German air forces pounded on their enemy and broke his will to counter attack; they destroyed his transportation facilities and made him immobile, starving for food and ammunition. In the seas, the German air forces sunk the enemy battleships once considered mightier and they made seas totally unsafe for the enemy. The siren of the German 'Stuka' dive bombers created enormous psychological impact and was deemed as a messenger of 'certain death.'

As the war went on, the power of air forces began to get prime importance among the war planners of various countries. This understanding led to the establishment of stronger air forces during the course of this war. An air force needs various types of aircraft, and hence, about 80 years ago, the aviation industry kick started to develop and build new aircraft for different roles. Along with the aviation industry, other industries that can play supplementary roles also started growing. These are still working in a fast pace to cope up with the ever changing doctrines.

In the Indian context, in the war with Chinese forces in the year 1962, had the Air Force been authorised to enter the war in the initial phases, many lives of soldiers could have been saved. Similarly in Kargil War, the intervening delay in bringing the aircraft to attack the enemy bunkers high up in the hill let many soldiers to sacrifice their life in capturing those bunkers. But when the use of air power started, it led the soldiers to literally walk over and occupy those positions without any fighting.

EMERGENCE OF MISSILE WARFARE

From time immemorial, man wanted something to do his job. In the case of war, man's desire to replace the pilot with a machine culminated in the development of guided weapons. Efforts to develop guided weapons were undertaken by war fighting nations of WW-II.

One such ingenious effort by USA during the fag end of WW-II, was to load aircraft with huge amount of explosives, fly and ram them against the difficult targets by remotely piloting the aircraft. The project was called as 'Aphrodite,' a top secret project in USA during WW-II. The results were not encouraging and so the project was discontinued. But it was not the case with Germany!

Germans had embarked on missile technology research from the 1930s and at the time the WW-II broke out, they had mastered the then state-of-art missile technology. It is only Germany which exploited technology to develop missiles as a weapon system. They possessed expertize to build various types of missiles.

They had Surface-to-Surface, Surface-to-Air, Anti-Tank, Anti-Ship,

Air-to-Air missile programmes. Unfortunately many of them even did not see production. Some had rudimentary guidance systems but many did not have any. This is because the semiconductor based transistor, an important component in electronics was yet to be invented by then. So, many of the missiles were unguided due to lack of technology. It will be interesting to understand the technological might of Germans that led to the creation of the following list of weapon systems.

“Rheinbote”, a long range artillery rocket; “Ruhrstahl X-7 (Rotkäppchen)”, wire guided anti-tank missile; Enzian (Gentian), an IR guided Surface-to-Air missile; “Rheintochter (Rhine Daughter)” and “Schmetterling (Butterfly)”, Surface-to-Air missiles with planned air-to-air variants; “Wasserfall (Waterfall)”, a Surface-to-Air variant of V2; “Feuerlilie (Fire Lily)”, Surface-to-Air Missile; “R4M rocket and Werfer-Granate 21”, unguided air-to-air rockets; “Ruhrstahl X-4”, wire guided air-to-air missile with anti-tank variants (X-7 and X-10); “Fritz X”, anti-ship guided bomb and “Henschel Hs 293” air launched anti-ship missile. Though these were tactical missiles the V-1 and V-2 were the teeth of offense and served as the baseline for the missiles that were to be deployed in the future.

V-1 MISSILE

The weapon was to be launched from specified ground sites through ramps consisting of a catapult.

It had a simple airframe with the Argus-built pulse jet engine mounted on the top of this airframe. The engine was simple in construction and pulsed at the rate of fifty times per second. These repeated actions gave the engine a distinct buzz and this distinct sound gave the name “buzz bomb”. The guidance system was based on a gyroscope for stability, a magnetic compass for sense heading, and a barometric altimeter for altitude control. A vane anemometer on the nose drove a mechanical counter which determined the distance. With all these systems, the missile used to fly towards the target with not much accuracy. The anemometer would count the distance travelled and upon reaching the target area a mechanism forced the wings to let the missile to perform a steep dive.

Production was accelerated across Germany and it was produced in mass numbers. By early 1944, V-1 was ready for action. The first missiles were launched on June 13, 1944, when around ten of the missiles were fired towards London thus the "V-1 blitz" in the lines of 'The Blitz (Lightning)' began.

Over 30,000 V-1s were produced during the war with around 10,000 fired at targets in Britain. Of these, only 2419 reached London killing 6184 people and injuring 17981. Though V-1 missile struck their target only 25% of the time, it was just a terror weapon and had little overall impact on the outcome of the war. But the V-2 missiles were a different story.

DEVELOPMENT OF V-2 MISSILES

One 'Captain Walter Dornberger' of German Army, was ordered to investigate the feasibility of rockets as a weapon in the year 1932. He was soon able to contact and recruit one young engineer called as 'Wernher von Braun', who is to be considered the Father of Ballistic Missiles in the years to come. Commencing work with a team of 80 engineers von Braun created a small rocket called as 'A2' in late 1934. Year later a rocket called as A4 was developed.

During 1942, Hitler authorized this A4 to be produced as a weapon and the missile was re-designated as V-2. The first V-2 attacks were launched against Paris and London on September 8, 1944. Subsequently, over the next eight months, a total of 3000 V-2 missiles were launched at allied cities including London and Paris. Due to the missile's ballistic trajectory and extreme speed, there was no defence against them.

The threats from V2 missiles were completely eliminated only when the Allied forces made the historic Normandy Landings and started advancing towards Berlin from the west. The Soviets, after shaking of their previous defeats, started advancing towards Berlin from the east.

The Americans took the captured V-2s to their White Sands Proving Ground, USA; and the Soviets took some V-2s to Kapustin Yar. From the basic V-2, the American and Soviet engineers developed ICBMs and satellite launch vehicles. Wernher von Braun can indisputably be called as the father of Ballistic Missiles.

AFTERMATH

Immediately after the end of the WW-II, the allied nations' camp split into USSR centric and USA centric. This resulted in a polarization of the world as Communist, Non-Communist and Non-Aligned blocks. Between the Communist and non-communist blocks, a silent war called Cold War began. To gain upper hand over the adversary both sides oversaw the development of deadlier nuclear weapons, enormous stockpiling, and development of intercontinental range bombers.

As the development of ballistic missiles with greater range and accuracy occurred, the slow and vulnerable bombers went out of the favour. This is because ballistic missiles were cost effective, capable of reaching greater ranges way out of the strategic bombers and apart from that they could reach the target in minutes. Above all they did not require putting a human being at the mercy of the enemy.

During sixties both sides realized that 'Ballistic Missile Defence (BMD) Systems' was a certainty in the near future. The seventies saw rudimentary ballistic missile defence systems being realized. So a missile system other than a ballistic missile was required. There is a military aviation saying, "either fly very high or fly very low." The very high ballistic missiles became vulnerable to the anti-ballistic missile systems. So the other option remaining was to fly very low. This very low flying could only be achieved by a cruise missile. Initial development of cruise missiles was for anti-ship missions pioneered by NPOM – our partners.

During WW-II and later, there was no technological maturity in guidance required for the high precision land attack cruise missiles. Due to these reasons thrust was not given for cruise missiles whereas ballistic missiles were favoured. But after realization of interception of ballistic missiles as possible, cruise missile was the only option to be taken up. Also by seventies and eighties, those guidance and propulsion technologies had got developed well in both sides of the world. This development enabled USSR and USA to concentrate on land attack cruise missiles. Many types of conventional air breathing missiles started to get fielded. The Tomahawk class by USA and Kh-55 class by USSR became dominant. Interestingly both resemble each other to an extent. Whatever weapon system the Soviets possessed, the Americans also had an equivalent resembling the Soviet's and vice versa. The espionage agencies of both the countries were so effective in getting the designs of what the other nation was working on.

A ballistic missile may carry heavier payloads. But a land attack cruise missile with a nuclear warhead was feared more, as a cruise missile is difficult to detect and intercept. By middle of the seventies, both USSR and USA had them up in their sleeves. These missiles were also able to carry non nuclear warheads and thus working as tactical land attack cruise missiles for precision strike. As the Soviet Union was dissolved, we did not get a chance to see the usage of the Kh-55 cruise missile. World's scientific-military community expected that cruise missiles will be fielded in the next major war. Americans did not disappoint for they used the Tomahawk missiles in the Persian Gulf War.

THE PERSIAN GULF WAR

The war was fought from 2nd August 1990 to 28th February 1991, between Iraq and a UN formed coalition force of 34 nations. The conduct of the Persian Gulf War has been serving a lesson for various nations regarding electronic warfare and air power. It was a war between a formidable air defence system and a formidable air force - a war between the doctrines of the past and the present. Above all it is a notable war in the annals of military history where the winning note was signed by precision strike cruise missiles. This war is like a primer for the conduct of future wars. A detailed description of the war is required for a clear understanding of the potentials of cruise missiles.

The Iraqi forces were very strong comprising of approximately 1.2 million ground troops, more than 5000 tanks, and 3500 artillery weapons. After occupying Kuwait they had well prepared with their defences along the Saudi-Kuwait border and Iraqi seashores.

To evict the Iraqi forces from Kuwait, the coalition ground forces had to face the Iraqi ground forces head on. But engaging Iraqi ground forces would have resulted in huge casualties for coalition ground forces. But that level of casualties would be unacceptable to the western countries on political grounds. Moreover, Saddam declared that Iraq would attack Israel, if Iraq was attacked. This would have forced Israel to retaliate against Iraq. Saddam's calculation was that such a situation would make other Arabian nations to withdraw from the coalition due to their dire enmity with Israel rather than Iraq. Withdrawal of the Arab nations from the coalition would have resulted in coalition forces being denied territories to launch military offensives against Iraq. So for the coalition forces, the war had to be won quickly before Iraq, with its known chemical warfare capability, could do any damage to Israel. Not only that, the war had to be won without escalating the war itself. Saddam played his cards in this fashion that either the entry of Israel will split the coalition or the western nations will succumb to their internal political situations due to mounting casualties.

To counter Saddam's strategy, the coalition decided to employ airpower to blunt the strength of the Iraqi forces. As a part of their strategy, the coalition forces wanted to disrupt Saddam Hussein from commanding his forces from his forty or more Command and Control(C2) centres situated throughout Iraq. A robust and redundant communication network is as essential as the command centre itself. This communication network survives with communication nodes like microwave relay towers and telephone exchanges. So these communication nodes were to be targeted.

Destruction of these nodes would destroy the network and it would be impossible for the higher commands either to receive situation reports or to pass orders. By destroying these command centres and communication nodes, orders could not be passed to the frontlines. The absence of orders will make the frontline defence to crumble due to lack of clear orders.

Next came in the list was Electricity Production and Oil Refining Facilities, Naval Forces and Port Facilities, Airbases to prevent Iraqi air operations. Then, came the NBC Weapons Research, Production, and Storage Facilities. After that, came the Scud missile production and storage and launch preparation facilities. Iraq had a stock pile of Scud missiles with which it could attack Israel. Next came in the list, the Railroads and Bridges that were the supply lines for Iraqi troops in Kuwait. Many of the telephone lines and fibre optic cables were also laid along these bridges and destruction of these bridges would serve two purposes: one is denial of transportation and the other is denial of communication link.

Saddam knew that the coalition forces will retort to air power. So he was ready for that with an ace in his sleeve – a robust air defence system. The coalition found that it was not easy to destroy the above said set of targets due to the presence of a gigantic Iraqi "Integrated Air Defence System" (IADS), called as the KARI standing up against the coalition air power. KARI served as an aerial fortress to protect this air space, a mammoth spider web over Iraqi Air Space. This system had virtually turned the Kuwaiti and Iraqi airspace as killing fields for coalition aircraft. The KARI consisted of lot of Air Defence Systems (ADS) like Anti-Aircraft Artillery (AAA) and Surface to Air missiles (SAM). These 'Air Defence Systems (ADS)' were complementing each other and providing mutually overlapping

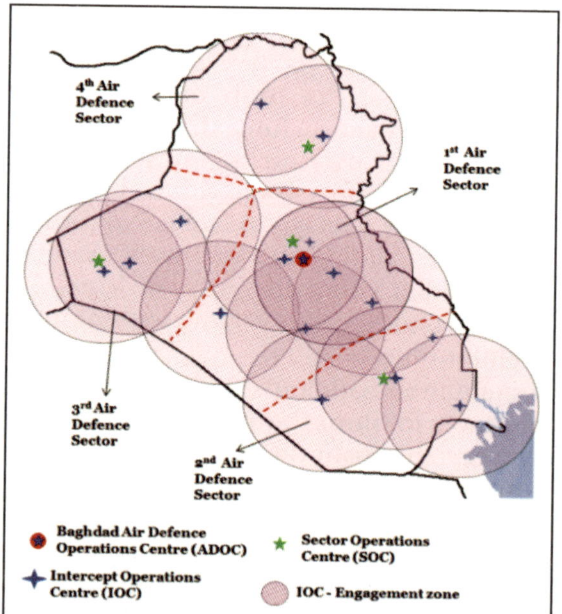

Typical Structure of KARI

coverage. Thus, an aircraft trying to destroy one ADS could be engaged by the neighbouring ADS.

The IADS was so constructed that neighbouring ADS belonged to different genre or of different operational frequency/modulation or different national origin. This arrangement was to degrade Electronic Warfare capabilities of the coalition forces. Jamming is an important feature of electronic warfare. Say, if the attacking aircraft tries to jam an ADS, the adjacent ADS will be unaffected by this jamming because of its entirely different nature. The adjacent ADS will be able to engage the attacking aircraft efficiently. Thus each ADS being different were able to protect the neighbouring ADS and as a cluster they were protecting the air space extremely efficiently. Above all the strength was each and every ADS was connected to a command centre called as Intercept Operations Centre (IOC). There were many Intercept Operations Centres, which were in turn connected with a Sector Operations Centre (SOC) that looked after the air defence operations of some two or three IOC. All these SOCs were connected to a central Air Defence Operations Centre (ADOC) situated at Baghdad under the direct control of Saddam. This ADOC was duplicated by around some 40 centres situated in Baghdad itself. The construction of those centres was such that even if anyone of the centre was destroyed, another centre could take over the operations without any degradation. The greatest advantage was such that without using its individual radar systems, an ADS could engage attacking aircraft. Because, once it turns on its radar, its location is revealed to the attacking aircraft, which could then destroy it. The brain of IADS was the networked commanding centres without which individual ADS had to operate on their own.

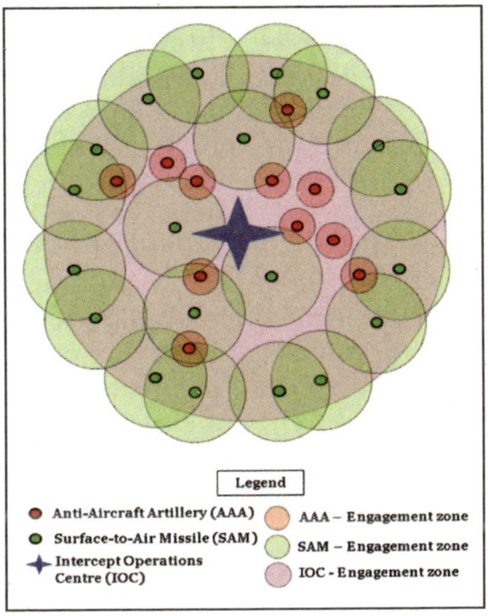

By the complication and threat level, this KARI itself became a primary target for the coalition forces. This KARI could only be attacked by aircraft but the irony is that the KARI was there to destroy the very same aircraft meant to attack KARI. So an alternate potent weapon other than aircraft was required. That weapon was the Tomahawk cruise missile.

The coalition forces struck their first blow at about 03:00 hrs local time on 17th January, 1991. The coalition aircraft took off to attack. But before this wave of attack, the Tomahawk cruise missiles were launched from US Navy's surface vessels and submarines. These missiles bypassed the air defence sites to hit the C3 nodes at very low altitudes. As many as 50 Tomahawk cruise missiles were fired against key military command and control posts, and government buildings.

And in the first few hours itself the KARI was almost degraded. And in the first few days KARI almost seized to exist. After this phase, the Coalition aircraft targeted the frontline troops in Kuwait, tanks, artillery, supply vehicles, bridges etc. Above all they destroyed the morale of Iraqi soldiers, who were deprived of their own aerial support, food, medicine and ammunition. After 40 days of continuous bombardment from the air, the coalition ground forces swung into action. The coalition army started from Saudi Arabia and directly engaged the Iraqi troops entrenched in Kuwait. But unknown to the Iraqi troops another force entered the great Syrian Desert and emerged behind the entrenched forces in what is called as left hook manoeuvre. They liberated Kuwait in less than 100 hours and the war was over.

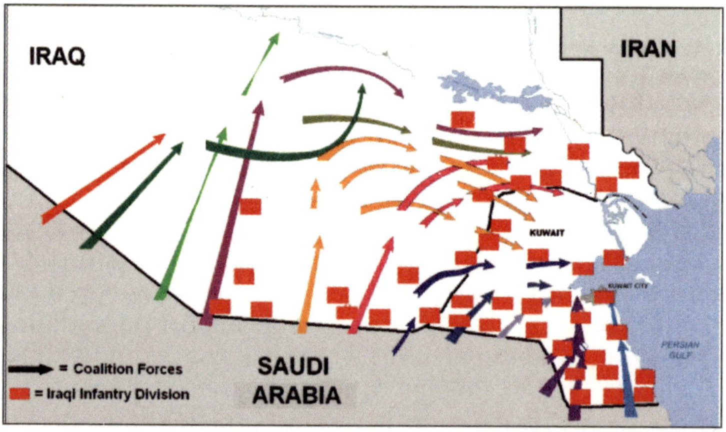

The Coalition Ground Force Movement (Left Hook)

But the key for this success was the near destruction of KARI through Tomahawk missiles. Had the 'Tomahawk' not been fielded, then the war would have been very difficult for the coalition forces due to prolonged conflict and loss of lives in greater numbers. From that war the effectiveness of Tomahawk class of cruise missiles was understood and thence has become a first strike weapon. Ultimately in total 1094 Tomahawk missiles were launched in all Gulf Wars and recently as many as 102 Tomahawk's were launched against Libya during the international intervention in Libya.

IN THE HIND SIGHT

This war gave a tremendous impetus for a paradigm shift in military thinking all over the world and especially in India. Also it was the time the international channels like BBC and CNN started to reach the masses through local television networks in India. So all the drama we were able to see live for the first time. The Scud and Patriot systems became hot topics for chitchats in office canteens, even on roadside teashops. People were mesmerised with the footage of 'patriots' intercepting 'scuds'. I can definitely say that awareness on the use of missiles came into the being in India.

After this war, our missile programs got great public attention. Those who didn't know about our missile programs were thrilled to know that India too was developing missiles. All this awareness resulted because of the penetration of television network and the news feeds on the war were enthralling. Who can forget those green tinted footages from night vision cameras showing the air defence guns blazing over the skies of Baghdad, along with the noise of air raid sirens?

At this juncture, I recall an event; during the first launch of SLV-3 on 10th August 1979, we even did not have colour television to monitor the rocket's second stage. For that stage the fuel was hydrazine with an oxidiser called as Red Fumic Nitric Acid (RFNA). During the countdown sequence, a certain procedure called as 'warm up sequence' involving the RFNA was executed. During the warm up, the valve that controlled the flow of the RFNA was opened for some seconds as per the procedure. The computer commanded the valve to close. But unknown to the computer the valve did not close, the RFNA started leaking. It had been leaking for almost 7-8 minutes. The computer while checking the various systems found that RFNA's presence was low and raised a flag asking to stop the launch. Now we had to give an override command as the RFNA's amount was well within the limits. So a launch command was given overriding the computer. That ill fated SLV-3 rose to the sky magnificently in a textbook performance. The first stage got jettisoned successfully to our cheers. Instead of further rising to the height, it slowly started dipping and then started diving in a swirling fashion then started spinning vertically and fell into the sea. Through our failure analysis we zeroed in on the leakage of RFNA and indeed we were watching that particular stage in a monochromatic (black & white) television. Had it been a colour television then ultimately we could have seen the red coloured acid leaking over the body of the rocket. We would have rectified the fault and would have saved the launch vehicle. What a growth in technology? Just in a decade we were able to see the live telecast of war and that too in colour. If in the common man's life, change occurs must faster, and then imagine how in a war every second may lead to a dramatic change.

During that Persian Gulf War, I was pursuing my course in the Harvard Business School, USA. I regularly used to look for the war's progress in the television and on that night of the attack a hunch made me to sit in front of the television. The initial media reports from Baghdad started coming in through CNN. The CNN reporter who was in Baghdad gave a report that he was able to hear the sound of aircraft and the sound of explosions but not able to see the aircraft. Immediately I understood that the sound he heard was the sound of the cruise missiles.

At the same time it struck me that in the event of a war the Indian armed forces are likely to face the absence of a credible precision attack cruise missile. So this was to be our magical first strike weapon. A cruise missile requirement was imminent. But that was the one of the busiest times in IGMDP. Prithvi was undergoing user evaluation in a rapid phase; the second flight of Agni TD was a failure from which we had to recover. We had to tackle the propulsion problems of Akash and the guidance related problems of Trishul. So at that point of time it became necessary to drop all futuristic ideas for the time being and push IGMDP forward. But nevertheless the quest for a cruise missile was on.

PART 3

The Voyage in the Uncharted Waters

Continuous effort - not strength or intelligence is the key to unlocking our potential.

- Winston Churchill

In this part, I am going to narrate about our voyage - mind to market journey. The voyage was in the uncharted waters because there was no trail for us to follow. BrahMos was the first JV for a military system in India. The JV was also the first of its kind to design, develop, produce and also market a futuristic supersonic missile through an inter-governmental agreement. To achieve our goals, we had to find our own path. Ultimately we found that and sailed on. This part contains the story of how we did that and many more incidents on the course of our voyage...

5

THE PIONEERING STEP: LEAP-FROGGING

The voyage in the uncharted water begins...

THE FASTEST MISSILE

When I returned back to India from the Harvard Business School in May 1991, the Persian Gulf War had almost ended. I made a detailed study of how Tomahawk missile played an impressive role in the Gulf War. This study was discussed by our missile team and we expressed our intention to go for the development of a cruise missile and ballistic missile defence system, as two important projects following IGMDP. My missile friends and I advocated the use of Akash as a BMD weapon for terminal stage engagement with feasibility report and the development of a possible cruise missile, both with the support of other countries. The collaborative model paved the way for several teams to visit and allowed us to analyze the BMD capabilities of other countries especially Russia, France and Israel. As IGMDP had delivered outstanding results at that time with many critical technology building blocks, other countries started believing in the technological strengths of India. Especially when there were Missile Technology Control Regime and technology denials, Indian scientists proved their capability, combating efficiently. Some technologically advanced countries believed us to progress with collaborative programmes for new developments in the missile programme.

The team associated with evolving cruise missile started looking at various options in terms of propulsion, guidance and control, seeker and configurations. The members took this task as an extra effort apart from their regular tasks. As per the traditions set for IGMDP, we wanted this new missile too to be a futuristic missile, best in its class at the time of deployment. But the question was how to make it futuristic? An answer came during one of our brainstorming sessions and it was to simply 'increase the speed'. In what way increased speed will make the missile futuristic? When the speed of attack increases, it automatically reduces the response time of the enemy. At that time all the operational cruise missiles in the lead countries, were flying only at subsonic speeds.

So, our team's motto became "Develop the fastest missile". At that time a fastest missile could be only a supersonic missile capable of travelling at speeds nearing mach 3. We carried out a meticulous study about the systems that would be required for the new missile say, the type of guidance system, type of missions the missile may carryout, the required airframe and most importantly the type of engine for attaining supersonic speed, etc. Also, side-by-side we analyzed the available expertise and the future expertises that we would accrue to develop the required systems for the missile.

From our analysis we concluded that we were confident in improving the guidance system of Prithvi missile for better accuracy and faster reaction time. Similarly, we were confident in developing the airframe. On the other hand, we required an engine belonging to a completely different class. From our studies we realized that the engine has to be an engineering marvel, an engine that has never been developed before. The said engine was a ramjet engine run by liquid fuel and which had the necessary features to propel the missile to speed of the order of mach 3.

We were competent in developing rocket engines, technically but at the same time, the thought came why supersonic cruise missiles are not operational in the world. Fighter aircraft flying at supersonic speed was realized but development of engine for missile application was a real challenge. At that time, we were concentrating on development of solid fuelled ramjet engine for Akash missiles. But even for developing jet engines (aircraft) there was a total blackout in technology and expertise. This incompetence became a road block. For this entirely new missile, we needed a ramjet engine whose technology had to be a derivative of the basic jet engines. Developing a ramjet engine at that point of time was itself a major program that would take at least 10 years.

The goal of IGMDP was to achieve self-sufficiency in missile technology that eliminated the possibility of buying the engine off the shelf. But even if we intend to do so, it was not known whether such an engine exists or not. Though the principles of IGMDP restricted buying, it did not prevent collaboration with another country possessing this ramjet engine technology. So the decision to collaborate was taken.

Naturally, a question may arise which nation would come forward to collaborate with us and what would be the win-win factor for both countries? To explain further, I have to tell you something about my ISRO days.

FRANCE - TECHNOLOGY COLLABORATION EXPERIENCE OF ISRO

I was working in ISRO at Thiruvananthapuram with Dr. Kalam for the SLV-3 programme in 1970s. One day, Dr. Sarabhai brought one visitor from France. He introduced the visitor as Dr. Curien, President of the 'Centre Nationale de Etudes Spatiales (CNES)', the French space research agency. He and Dr. Sarabhai helped to address some issues in the design of the fourth stage of SLV-3 for which Dr. Kalam was the Project Manager.

Dr. Curien then enquired whether the fourth stage of SLV-3 be utilised as a stage for their *Diamont* rocket that they were developing. Sarabhai who was witnessing the conversation gave a smile of success! Dr. Curien's enquiry was nothing but an accreditation for an "infant space programme" that it could deliver wonderful results. Unfortunately, the French cancelled the *Diamont* programme and communicated that our design was not required.

Some years later after this incident, France was designing a new launch vehicle called *Ariane*, a successful design indeed. You might have read news in newspapers about the INSAT being successfully stationed in the orbit by an Ariane rocket launched from '*Kourou*' of *French Guyana.* The *Ariane* rocket is a real workhorse of France and has placed number of satellites in orbits. Ariane was propelled by an engine called Viking.

There was an agreement on transfer of Viking engine technology, signed between ISRO and CNES in the year 1974. Through this agreement, ISRO scientists were allowed to work in the French propulsion systems lab *Société Européenne de Propulsion* (SEP). The agreement scheme was developed in such a way that the ISRO had to provide 100 man years to develop a version of Viking engine and will also receive the engine technology. France at that time badly needed a highly trained and knowledgeable workforce to quickly complete their Ariane programme. Their deadlines were crucial. Due to this reason ISRO had to provide 100 man years. Against this obligation of 100 man years, there was no restriction on Indian scientists to learn about the engine technology. Our scientists, who worked there, absorbed the technology within five years. Not only did they attain the status of 'know how' but they also acquired a status of 'know why'. When they came back to India, they had already mastered engine technology with which our engine known as "Vikas" was developed. It is this Vikas, which propels our PSLV and GSLV rockets.

INDO RUSSIAN RELATIONSHIP

The Indo-Soviet/Russian relationship has been flourishing for more than half a century. The relationship, which is not a military alliance, exists almost in all spheres like space research, cultural exchanges, economic cooperation, military weapon system sales, mutual politico diplomacy, etc.

The relationship blossomed during the end of year 1955. Soviet leaders 'Nikita Khrushchev' and 'Nikolai Bulganin' visited India and from then onwards, the relationship kick started between the two countries. USSR diplomatically supported India during the wars with China and Pakistan in the years 1962, 1965 respectively. When war clouds were looming during the year 1971, the Chinese expressed their deliberate support to Pakistan. USA and Pakistan had strong ties and Pakistan was a member of CENTO and SEATO military pacts. Also USA and China had also started building relations against USSR.

In such a situation India and USSR signed an important agreement called "Treaty of Peace, Friendship and Cooperation," on August 1971. According to this treaty, when one of the two nations is at war, the other should render military support. The treaty emphasised mutual respect and sovereignty for each other irrespective of the ideological differences. It was legally signed for a period of twenty years as Indo-Soviet cooperation. Just because of this treaty, other countries did not have the nerve to intervene. Initially known as the USSR, and later the Russian Federation, has been a consistent supporter to India on many crucial issues.

Relationship in Space Research

We have a deep rooted relationship with Russia in space research. The Russian contribution in the development of Indian space research capabilities has been remarkable. Our first satellite *'Aryabhatta'* and the remote sensing satellites *'Bhaskara-1'* and *'Bhaskara-2'* were launched through launch vehicles of the Soviet Union. Our operational remote sensing satellites IRS-1A, IRS-1B and IRS-1C were also launched through Soviet launch vehicles.

In the early 80s, a joint manned space program was planned between

ISRO and Soviet Union's Intercosmos. Three Cosmonauts, Yuri Malyshev, Gennadi Strekalov from Russia and Rakesh Sharma from India were selected for the manned space programme. On 3rd April 1984, Soviet launch vehicle 'Soyuz T-11' took off from the Baikonur cosmodrome to the 'Salyut-7' - a low earth orbit space station. He spent a total of 7 days 21 hours and 40 minutes in space aboard Salyut-7, conducting an earth observation programme concentrating over India. Mrs. Indira Gandhi asked Sharma during a teleconference, "How India was looking like?" and Sharma answered, "Saare Jahan Se Achcha" (better than all lands) and made millions of Indians proud. The next major cooperation was the Cryogenic engine for GSLV as its third stage. Russia provided the Cryogenic engine with stage for the initial GSLV flights. The agreement to develop an engine in India, based on Russian design was stopped by USA, making it a MTCR issue. ISRO has successfully developed its own cryogenic engine for the future flights.

Economic Cooperation

The economic cooperation between the then USSR and India emerged during the early fifties. Initially it came as Soviet help for the Bhilai steel plant, one of the first of its kind in India. This installation helped to kick start the iron and steel industry development in India. Similarly, considerable aid was provided by USSR for the Indian Institute of Technology Bombay. Soviet cooperation was extended over many strategic civilian production areas like heavy engineering machinery, oil exploration and refining, coal mining and power generation etc. These cooperation in due course led India to become self sufficient in those areas.

The prime root of the economic cooperation between the two nations was the Rupee-Rouble cooperation. USSR provided weapon and equipment to India at nominal interest rates and the facility was that the repayment could be made in kind and not in dollars or roubles, on a long term basis.

Cultural Exchanges

The success of BrahMos is only because of the cooperation, which in turn is a result of the goodwill of both nations towards each other. In

India, after the 1971 war, USSR got a special place in the minds of Indians. Whenever a Soviet leader or a Soviet citizen visited India, they were welcomed wholeheartedly by the people of India. Soviet names like Valentina Tereshkova (first female cosmonaut to reach space), Yury Gagarin (first cosmonaut to reach space), Sergey Bubka (famous pole vault athlete) and many more became household names.

Similarly, India and Indians had a special place in the heart of Russians. I would like to quote one particular incident. During our initial negotiation phase for BrahMos, we were in Russia. We were travelling in a metro. A man who was sitting on the opposite seat saw us and with a smile asked us "Raj Kapoor?" He mentioned the famous Indian actor. We were startled and later came to know that Russians were great fans of Raj Kapoor. He talked with bitterness about the breakup of Soviet Union and the lost glory. When the train reached the station he had to get down, he wished us good luck and parted saying, "Indians are our friends and will always be."

Goodwill and the positive contact between individuals are the success foundation of technical cooperation in space and defence.

Arms Trade

USSR and then later Russian Federation has been our major source to procure weapon platforms when no other nation was willing to sell to us. During Soviet Union times, many weapon systems were sold to us on loan with cheap interests or for a very friendly price. After the dissolution of USSR, we were able to get many systems that were India specific with a Russian friendly price. For Russians, since it was India, they agreed to sell many modern systems including the much notable Sukoi-30MKI aircraft with many western avionics. Similarly, six state-of-the-art stealth frigates called as "Talwar class" have been constructed in the Yantar Yard at Kaliningrad. Also, as a gesture of friendship, a nuclear powered submarine was also leased through which India got to know operating such complex vessels.

Integrated Long Term Programme (ILTP)

During the world wars and later the cold war period, Russia had established many design bureaus which supported military industry complex by providing research and development effort in high technological areas. Over a period Russia had a knowledge pool at different establishments looking for opportunities to carryout research in newer areas. India and Russia decided together to introduce Integrated Long Term Programme (ILTP) to enable continuation of research work and sharing of knowledge. DRDO utlilised this opportunity and formed a joint R&D Group for initiating research cooperation in selected areas and for providing consultancy in complex technology development projects. Many civilian, space and military projects have benefited from such cooperation.

One of the leading design bureaus of Russia is the NPO Mashinostroyenia that specialises in launch vehicle and cruise missile technologies. Let me narrate the highlights of the organization in the succeeding chapter.

6

NPO Mashinostroyenia is a legendary organization of Soviet Union established by Vladimir Nikolaevich Chelomey with a large reservoir of knowledge in missile technology especially in cruise missiles. So in this chapter, I am going to narrate about NPOM.

VENTURING PARTNERS' STRENGTH

NPOM is the acronym of *"Naochno Proezvodsvyenoye Obyedeenyeneeye Mashinostroyenia"* meaning *"Scientific and Production Machine Building Association"*. "NPOM" is one of the leading space and rocketry companies of Russia that has realized more than 50 large scale missile and rocket programmes. Apart from this, it had embarked in three national defence programs: Arming the Soviet/Russian Navy with cruise missile complexes capable to be launched from surface, underwater and land based platforms; arming the strategic nuclear forces of USSR with intercontinental ballistic missiles; Development of space systems and spacecraft, automatic and manned orbital stations in the interests of USSR.

NPOM has been a specialist organization for cruise missiles and around 60-70 per cent of the cruise missiles that are being operated by the Russian navy were designed and developed by NPOM. Though NPOM is presently an open joint stock company, it was started as a design bureau, in the year 1944, by a legendary scientist and engineer called, 'Vladimir Nikolaevitch Chelomey'. He can be considered equivalent to Vikram Sarabhai. We will briefly look into the interesting history of Chelomey and his association with NPOM.

Vladimir Nikolaevich Chelomey was born during the WW-I. After his doctoral research he was concentrating in pulsating jet engines and related dynamics problems in Moscow. His research was in a time when the WW-II was on full swing and Hitler was creating havoc with his V-1 cruise missiles barrage against London. After great difficulty the Britons equipped themselves to shoot V-1. Since Britain was an ally of USSR during WW-II, some of the remains of V-1 were sent to Moscow.

In the year 1944, Chelomey was asked to reverse engineer the German weapon. He was a visionary who looked beyond decades and he had rightly understood the potentials of cruise missiles. Finally, Chelomey was appointed as chief designer of the aviation plant No. 51. By the middle of 1945 he had designed the first cruise missile of USSR. Though it was just a copy and was not produced in mass, but it definitely marked the beginning of the birth of cruise missiles in USSR.

Chelomey proposed improved versions of his air launched cruise missiles called as flying bombs like 10X, 12X, 14X and 16X to be launched from long-range bombers, to the Soviet Air Force. A Special Design Group (SKG-10) was created in the town of Tushino, near Moscow from where Chelomey started promoting his winged missiles for naval use, particularly for submarines. Soviet Navy was also interested in such projects, as the Soviet strategy was to counter the aircraft carrier based western naval doctrine with submarines equipped with missiles.

By the year 1955, the SKG-10 was converted as *"Opitno Konstroktorskogo Bureau – 52"* meaning Experimental Design Bureau-52 and simply OKB-52. Post WW-II era, aircraft carriers became the principal stay in naval power. The cold war between USSR and the west started. Attack against USSR by NATO aircraft taking off from aircraft carriers was perceived as a significant threat. USSR was in a position to find effective ways to counter the naval threat and was trying to identify a suitable weapon - a deterrent against the navy of the west. The weapon was supposed to have long range, higher hit accuracy with ample destructive effect, in flight manoeuvres and to be operationally capable during all weather conditions.

After various analyses, cruise missiles launched from surface ships and especially launched from submarines were identified to possess all the desired characteristics.

Initially, cruise missiles launched from submarines were given the role of land attack with a nuclear payload while anti-ship attack was considered secondary. Changes in the Soviet doctrines and the need for cruise missile were felt in the early fifties and by this time NPOM got established as OKB-52.

With hands-on experiences gained by the development of the 10X-series air-launched cruise missiles, Chelomey took part with his design in a competition that was held for the design of the new cruise missile. His design was ingenious and revolutionary which included the missile's wings being folded during storage. When the wings are folded the body of the missile becomes almost cylindrical and can be stored in tubes, technically called as canisters. When launched from canister, the wings get unfolded in an ordered sequence in the air. Chelomey got this revolutionary idea by watching the birds coming out of a small hole in the tree and spreading its wings the instant it came out. Ultimately, this design won the competition in which other famous designers also took part. From that moment, folding wings and container launched configuration became the inherent part of the NPOM cruise missiles. This missile was named as Project-5 and got inducted in the Soviet diesel electric submarines.

The missile's design was a pencil shaped fuselage with swept wings and propelled by a turbojet engine with an air intake under the belly. When launched, the missile used its Rocket Assisted Take Off (RATO) boosters for launch. After reaching the desired speed, the used boosters would jettison and the turbojet engine would start. This feature also became the standard feature in the forthcoming NPOM cruise missiles. The P-5 had a range of 500 kilometres and the guidance was based on autonomous inertial navigation system. Even with limited accuracy of guidance systems at that time, the P-5 was usually armed with a nuclear warhead.

Immediately after induction, the same missile was equipped with a radar navigation unit and was named as P-5D. It was inducted in service by the year 1962, for anti-shipping roles. The very same P-5D was also produced as a surface-to-surface missile complex for tactical use, as a coastal defence missile. This variant was called as C-5. During the development processes, most complicated scientific-technical problems kept evolving. Launching the missile from the container and algorithm for designs of wings unfold was too complex a job. A millisecond miss in synchronization would eventually lead to catastrophe. Chelomey himself being a professor formed a close cooperation with academia and NPOM and solved all the technical issues. When a team from NPOM visited DRDO during the formation of JV and they came to know about the networking of academia and industries for our IGMDP, they were highly impressed, since it was Chelomey's style of working.

After the P-5 programmes, Soviet Navy was relieved from the task of land attack and it was assigned to the strategic rocket forces for which Chelomey had already contributed. By this time, NPOM was working on two anti-ship missile variants for dealing with western aircraft carrier task forces. One variant called as the P-35, was capable of being launched from surface ships and the other named P-6 was capable of being launched from submarines. A land based coastal defence variant of P-35 was also developed. These missiles had INS based navigation system and an active Doppler radar based guidance for the terminal attack sequence.

The missile was way ahead and advanced from its predecessors both in terms of technology and capability. But it still required the submarine to stay on the water surface - its weakest attribute. So a new missile called as "Ametist" was conceived which had to be capable of being launched by submarine, in submerged stage. This missile became the world's first cruise missile capable of being launched from a submerged submarine. Now, the Soviet Union was far ahead, at least by a decade, to the west in acquiring such technology.

The Ametist was followed by an improved missile called as Malakhit. This missile had multiple launch capability which means that the same missile can be launched from a submerged submarine as well as from surface ships. Malakhit possessed longer range, improved guidance systems and was loaded with superior flight profile than its predecessors. Eventually, this missile got inducted by the Soviet Navy in the early 1970s. The missile had range of around 70 km from a submarine or 110 km from a surface vessel. It could carry 500 kg explosive warhead or a nuclear warhead and flew at a speed of 0.9 mach.

After these missiles, NPOM began to develop a new missile complex called as P-500 "Bazalt". This missile was a replacement for the P-6 and P-35 missiles and had a greater range and speed with improved flight characteristics. This missile was the first cruise missile capable of travelling at a supersonic speed of Mach 2 which was later taken over by our BRAHMOS. This missile was inducted in the late seventies and early eighties. Till that date, Bazalt used to be the most superior system when compared to any other missile of those times.

Even during the days of designing the Ametist and the Malakhit missiles, Chelomey concluded that it was necessary to design long-range missiles with universal launch capabilities. This line of thought resulted in the birth of a new missile called as P-700 Granit, designed in the early 1970s to replace the Ametist and the Malakhit. But to fit the new bill, the missile needed a robust system that could provide information about the targets. This requirement could be sufficed only through sea surveillance satellites and hence Chelomey began to develop such a system. It was assumed that sea surveillance satellites would be brought to the orbit by launch vehicles developed by NPOM. The additional role of this launch vehicle was to deliver the world's first manoeuvrable satellite to space, whose development was under way at NPOM. A launch vehicle meant for these purposes was developed and designated as UR-200.

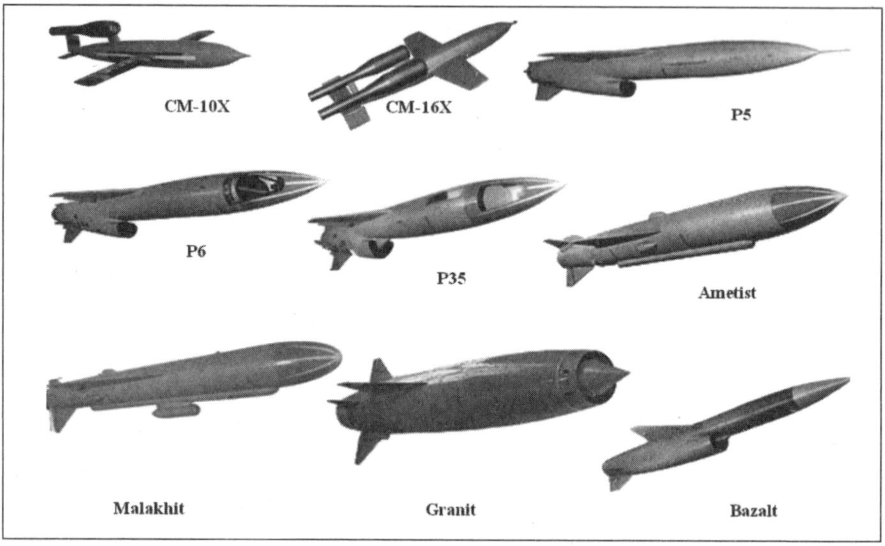

NPOM AND SPACE VEHICLES

When manned space programme were conceived in USSR, launch vehicles weighing around 500 tonnes were needed. NPOM designed and developed a launch vehicle called as 'UR-500' in the 1960s. Also the experiences acquired during the development of UR-500 allowed NPOM to design a launch vehicle "Proton" for the Soviet lunar program. The Proton launch-vehicle created history by ensuring successful launches of many manned space stations like "Salut" family and automated stations like Almaz, Venus, Zond, Luna and Mars. These sections formed the basic compartment and other modules of the famous "Mir" complex. The Versatility of UR-500 and another project called UR-200 allowed them to be used as ballistic missiles and not just only as launch vehicles.

These experiences led to the development and commissioning of canisterized intercontinental ballistic missile UR-100. Till the middle of 1980s, NPOM developed and put into service several versions of missile complexes with ICBM of this class (UR-100 K, UR-100 У, UR-100N, UR-100N U TTX). NPOM has been a dynamic organization with specialisation in cruise missiles and space vehicle development. They always used to start new & ambitious project when another project was nearing completion. NPOM had designed and developed various spacecraft and launch vehicles. Six of those systems are now on combat duty in Russia.

Owing to the ever dynamic Warfield they wanted to develop a new cruise missile for the Russian Naval Fleet that could be launched from underwater as well as from the surface as anti ship missile. Previously, all the cruise missiles of NPOM had turbojet propulsion due to which they could only attain subsonic speed. They envisioned that new missile would be designed with a ramjet engine based on liquid fuel – a quantum leap from the previous missiles. NPOM eventually succeeded in the development of unique ramjet engine equivalent of which does not exist anywhere in the world. NPOM conducted several ground and flight tests to prove the engine of a supersonic anti-ship cruise missile. The disintegration of the Soviet Union in the early 1990s affected the economy of the Russian Federation which ultimately stopped all the Research & Development projects. Thus, NPOM could not progress further on this project.

Any research oriented organization can withstand number of failures but the negation of a project at its completion stage, was a fatal blow. We have also faced many such situations in DRDO. Every organization strives to keep its projects live by any means, and NPOM was also one of them.

NPOM's contribution for India's Missile Projects

During the development of Surface to Air Missile Akash, we had to repeatedly encounter failure during the booster-sustainer transition. In spite of many efforts to pool the best scientific minds from the country, we were unable to solve the problem to achieve perfect transition required to fly Akash at supersonic speed. Therefore, we approached NPOM for providing consultancy services to resolve the issues. NPOM came out with a descriptive analysis and suggested tests and methods to overcome the problem. We also had consultancies on many other areas in missile technology for DRDO. Mutual understanding among the technical experts of DRDO and NPOM led to confidence building and envisioning of future.

In India, we were in the process of identifying various technologies required for the realization of a supersonic cruise missile. The thought we had in mind was to collaborate in advanced technologies so that the development time can be minimized. So, when we came to know about NPOM's efforts we perceived that it could be the right organization, and that the partnership may lead to complement the capabilities of DRDO, that would result in win-win situation.

7

FORGING ALLIANCE: GENESIS OF THE JOINT VENTURE

The story of the formation of JV is really pretty long. The number of man-hours spent in negotiations and discussions, and the amount of paper communication that were carried out are countless. This single story of establishing the JV Company against all odds itself can be written as a separate book, so I am giving this story in a nutshell.

TRACING BACK IN TIME - AN ANONYMOUS COMPANY

You might have heard and read about various top secret projects around the world. I am absolute sure that many of you might have longed for a top-secret project in India also. For your knowledge, BRAHMOS missile programme was a top secret programme to start with.

On 05th Dec of 1995, a company was registered with a name BrahMos Private Limited at Delhi. All the formalities were completed in the Ministry of Company Affairs. Even in that office, nobody knew what our company was made for. It was gearing up period for the Indian economy spurred by which lot of first generation entrepreneurs started their business. So this newly formed company was just another company among hundreds of registered private companies. But this gave us an edge by giving cover for the purpose of the secrecy, right from the very beginning.

USSR was once called as an 'iron curtain' since the activities prevailing in the country were not revealed outside and secrecy was the mantra. Maintaining secrecy did pay well for USSR in upholding military parity with the west. Nowadays, you can find number of resources over the internet giving data regarding the Russian weapons and systems. But, in the past, existence of a particular weapon was itself a secret. Western intelligence agencies were battling hard to get data regarding weapons.

In the early nineties, USSR was officially dissolved and Russian Federation emerged as the successor nation. Those were the years of trouble for both India and Russia. Russia was badly affected by financial crisis due to the sudden dissolution. Geopolitically, we were lacking strong friends to support our noble and genuine concerns. Our defence forces suffered with lack of spares for their weapon platforms since 90 per cent of them were of Soviet origin. Many of the defence industries which were situated all over the Soviet Union went along with those breakaway nations. Slowly, Russia started regaining its glorious past which has brought happiness to India. During those difficult times, DRDO helped many Russian design bureaus to continue their R&D projects to preserve those Russian industries with work orders, so that the scientific talent could be preserved and the exodus of the scientists could be controlled. I was heading the R&D group from the Indian side to maintain the collaborative efforts.

During one of the consultancy activities at Moscow, I was discussing with our Russian friends about the outcome of the Persian Gulf War. The conversation slowly moved towards the conduct of the war and the usage of air power. Inevitably, the focus shifted towards the Tomahawk and the role played by it. I expressed my wish for the first time that India too should have a cruise missile but superior to that of Tomahawk. The NPOM team got interested and the discussion turned towards our search for a liquid ramjet engine. Then suddenly the NPOM team consulted themselves in Russian language in a low voice. One of the team members told that they already possessed a fully developed liquid ramjet engine which can give supersonic speed. Then he told the story of their cruise missile development programs and about Dr. Chelomey's contributions. The gist of whatever he told I have already narrated earlier. He ended his narrations with their venture developing seven types of cruise missiles for Russian Navy, all for anti-ship applications. He told that they had test flown a prototype missile with the liquid ramjet engine but refused to divulge any more information. He regretted that they could not continue the programme due to severe economic crisis in which the development activities were stopped in all design bureaus and in all spheres. I pondered over this discussion for many days after coming back to India.

In my next visit to NPOM in 1993, I met Dr. Yefremov, the then Director General of NPOM. Dr. Yefremov recalled the development effort put forward by NPOM in developing the supersonic engine which according to him was very unique in the world. He disclosed that the Germans wanted to acquire the technology from Russia and it was not agreed to. However, India being a close friend of Russia, it would be possible for Russia to agree for technology transfer and produce these engines in India. This would enable India to not only manufacture the engines but also configure a supersonic cruise missile. He then told Mr. Strakhov and Mr. Khomyakov to show the engine to me and explain its significance. I was taken to a hangar in the afternoon to see the engine and all other things were covered so that the secrecy was maintained. When I heard the explanation about the engine, I realized the importance.

Dr. Yefremov continued the discussions with Mr. R. Ramanathan the then Addl. Financial Advisor of DRDO, and V. Adm (Retd) Bharat Bushan, who was the Director General of ATVP when they visited NPOM. Based on this discussion, Mr. Ramanathan generated a note (an official communication) on the possibility of transfer of technology of the engine and the need to involve a team for details. Further, deliberations led to the idea of forming a collaborative effort to design and develop a supersonic cruise missile. A joint feasibility study team was set up with me, Mr. Ramanathan and Mr.

Venugopalan from DRDO side and Dr. Leonov, who is the present Director General of NPOM; Mr. Strakhov and Mr. Khromoshkin, from the NPOM side to evolve the collaborative methodology. This team during the deliberations found out that this engine could be used with a particular configuration for the missile and substantial inputs were available for realizing a supersonic cruise missile if we can put together the technological assets from both the countries. We discussed the above details with Mr. Ronen Sen, the then Indian Ambassador at Russia who suggested that instead of collaborative efforts, it could be a joint venture. He then wrote a letter to Dr. Kalam advising him to work out the joint venture format with independent status. The scope of the committee then became to set up a joint venture to design, develop, manufacture and market the advanced missile. This involved technology sharing, work share, development effort, schedule, costs and study of market potential.

The preliminary analysis of the proceedings of the joint committee resulted in generating a proposal from DRDO to the Government of India recommending a joint venture concept and seeking in principle approval. Mr. P.V. Narasimha Rao, the then Prime Minister who was also the Defence Minister reviewed the proposal and approved the idea to go for a joint venture.

The three member team comprising Ramanathan, Venugopalan and myself were debating on the formation of the joint venture particularly when the Soviet Union disintegrated into number of countries, the missile experts of Soviet Union had to leave the design bureaus and chaos prevailed everywhere due to poor economical status. NPOM was not in a position to talk of share capital as Russia was gripped by a tough and adverse economical situation. NPOM therefore came out with an idea that they will provide knowledge and technology as their share to be valued as 50 per cent of the total capital. The remaining 50 per cent required for the full development would come as hot currency from the Indian Government. The original estimate for the joint venture was \$1500 million and \$750 million to be contributed by India. This was not accepted by the Indian side. Detailed analysis and India's insistence over Russian contribution in cash led to scheduling of several meetings of the committee and the work progressed slowly. Tremendous amount of perseverance and frequent visits to Russia finally led the committee to settle down to $250 million for the development effort of a shore based anti ship missile complex.

NPOM discussed the issue with the Russian Government and coaxed them to address the financial problems. The political climate was also not conducive and governments were not that stable to arrive at a concrete decision regarding such policies. Here, I have all praise for Dr. Yefremov and great respect for his efforts to get approval for the JV Company from the Russian Government through his tireless persuasion. The Russian Government then appointed a specialist committee to review the capabilities and technical competencies of DRDO to enter into a joint venture. Now, the debate started inside DRDO and we selected specific establishments to show case the technological competence and milestones achieved. The specialists committee from Russian Government led by Mr. Sepnov, Head of Russian Military Industries and Dr. Yefremov made a visit to DRDO establishments at Delhi, Dehradun, Hyderabad and Bangalore. The Russian specialists were awestruck and very happy on seeing our developmental capabilities especially in missile technology, guidance and control, software packages, computing capabilities and electronics systems. Their confidence level shot up to high levels. The specialists committee went back with full of enthusiasm and their report suggested that DRDO as the right partner for the JV company.

Again financial concerns from NPOM rose since they still were not able to arrange funds for the JV Company. Now NPOM insisted that their process stock in terms of liquid ramjet engine, advanced booster, seeker, the design knowledge and expertise to be taken as the share capital from their side as they had spent huge amount to develop these technologies. Moreover, in the prevailing economic situation, there is no money available. Mr. Ramanathan had a firm view that joint venture can be evolved only if both sides can contribute money in cash as investment. It became a stalemate. We were debating at the Embassy with Mr. Lambah who succeeded Mr. Ronen Sen as Ambassador. He gave the idea that the debt repayment being made by India to Russia every year for the credit can be a possible option to provide as a share capital from the Russian side. This audacious idea took everybody by surprise and enabled us to exhale a sigh of relief. But, it is necessary for both Governments to agree that repayment in kind from India to be converted to cash and invested over India. Mr. Ramanathan and I effectively canvassed this idea to the Indian side to agree to this proposal. After several discussions in the Finance Ministry and Reserve Bank of India, we could see the light of hope from the Indian side. But the big question was whether the Finance

Experts from Russia will agree?

After finding the way of converting the money India had to pay against the debts to Russia, as an investment, Mr. Ramanathan and I went to the Russian Finance Ministry along with Mr. Khromoushkin and Mr. Semaev, the representatives of NPOM. There, we met a lady who was a top Russian bureaucrat and presented the JV proposal to her. She was of typical Russian built, clearly more than a foot taller than us and had a very strong voice. Because of the presence of Ramanathan, I was confident that we would be able to handle that lady, as generally scientists and bureaucrats don't generally gel with each other.

She was puzzled by the proposal for JV at that time because of the tight grip of economic crisis over Russia. She told that the proposal was a funny idea and asked us politely to clear the place saying that Russia's financial conditions were not conducive for undertaking a JV. When we emphasized about the conversion of the debt money as an investment she started laughing and asked us in what days we were.

She said,

"It was Soviet Union but now it is Russian Federation.

It was friendship then. Hence, the Rupee-Rouble trade adjustments were made.

But now for partnership, Dollar is the currency.

How it is possible for me to take the debt money as an investment?"

We were a bit taken back and discussed the issue with Mr. Lambah, the then Indian Ambassador for Russia. Mr. Lambah handled the financing issues remarkably well. He took me to Mr. Kudrin, the then Russian Deputy Minister for Finance and spoke about the JV for which the minister saw a new ray of hope and replied in full enthusiasm. He gave his acceptance and support for the project then and there. For them a JV was a novel idea and a success of such joint ventures would be a way out from the economic turmoil. After getting the nod from him, the Ambassador asked me to meet the lady in-charge of financial arrangements from the Russian side.

Again me and Ramanathan knocked at the doors of that lady and told that the Deputy Minister for Finance was willing for the JV. But she again gave a cold shoulder stating that she had no instructions from the ministry and she was not agreeing for any further discussions. So we returned.

A few weeks later, Dr. Yefremov met the Deputy Minister for Finance, the outcome of which made him deliver the necessary instructions to the ministry officials. I again met the lady official in the Finance ministry knowing well that she has got all the directions. When I approached her this time, she was very friendly and stated that she was making a document exploring the possibilities for JV. In truth, she had no opposition for the proposal but was a staunch follower of rules and instructions. When she got the green signal, she rendered an invaluable cooperation for the project. Her proposal was sent to the then President of Russian Federation Mr. Yeltsin. Those papers had travelled for two months crossing so many bureaucratic procedures. But when the papers reached the president's table he suffered a heart attack, not because of the proposal, of course!

The proposal of Joint Venture was discussed between the Governments of Russian Federation and India. The idea of the two governments was to form a JV with three shareholders holding shares: DRDO (49%), NPOM (49%) and an Indian financial institution (2%). This will provide equal partnership to DRDO and NPOM, and at the same time makes it 51% for India to have a JV Company instituted in India but with 49% for DRDO, the JV will become a private sector company. The concept of private sector company was agreed to by both the Governments in order to give flexibility to the organization which is controlled by both the sides and not imposing the regulations of both the Governments in the day to day affairs, to ensure a rapid progress. Now, identifying a third partner became an issue since the joint venture was classified as a top secret project. There were concerns that inclusion of a financial institution may compromise the secrecy. So the financial institution was eliminated and with only DRDO and NPOM as the partners holding 51% and 49% each. Mr. Ahluwalia, the Finance Secretary reviewed the proposal and raised the question that this ratio may result in losing the advantage of private sector status for the joint venture. Debate started in the Finance Ministry and consultants and Mr. Ramanathan made a strong recommendation to make it to 50.5% and 49.5%. Mr. Ahluwalia again wanted the clarification whether 50.5% will give an advantage of Private Sector or will it be treated as PSU. The Ministry of Company Affairs confirmed that though the Government of India holding majority share of 50.5%, the Company will still be private sector and it has to be registered with the Registrar of Companies.

In India, when the government share is 51%, then the Company will be a Public Sector Undertaking (PSU) like Hindustan Aeronautics Limited (HAL), Bharat Dynamics Limited (BDL). This places it to be under the direct control of the Indian Defence Ministry which was not acceptable to the Russian side. They feared that procedural formalities and government controls from the Indian side will delay the operations of the joint venture. So the best way to tackle this difficulty was to make the company a private sector with less than 51% for the Indian side. The Russian side demanded equal partnership and control in the operations of the company. In a novel way, we did that just by evolving a new share percentage of 50.5 for DRDO and 49.5 for NPOM.

The funding pattern for the joint venture was presented to Mr. P.V Narashimha Rao, the then Prime Minster holding Defence portfolio also, by Dr. Kalam, Ramanathan and I. He went through the intricacies of the formation of the joint venture, financing scheme and asked a few questions accompanied with his usual facial gesture which could not be understood by anyone. Finally, he nodded his head and said, **"It is a good idea and if it clicks it has a great future"** - a bold and foresighted decision on his part. The political will exhibited by him gave tremendous confidence that laid the foundation for the successful emergence of this venture. The Government approval was then translated by registering the BrahMos JV at Registrar of Companies at New Delhi, on 5th December 1995. The approval from the Russian side was yet to come.

The year 1996 and 1997 saw much political turmoil in Russia. Hence, the proposal had to go through many hands again and again due to which we lost two very valuable years. Myself and Mr. Pukhamov, the then Deputy Minister of Defence, Russia finalised the draft of the Inter Governmental Agreement (IGA) with the participation of NPOM team. By that time, the draft for the IGA was prepared. Finally, we got the decree issued by the President of Russian Federation permitting the JV. On 12th February, 1998, the Inter Governmental Agreement was signed by Dr. Kalam on behalf of Government of India and Mr. NV Michaelov, Deputy Minister of Defence on behalf of

Historic signing of the IGA by Dr. Kalam & Mr. Michaelov

Russian Government at Moscow.

Again after a series of persuasion, the Russian Government issued a decree on the modality of financing the JV on 12th August 1998. This presidential decree gave clearance for NPOM as it accepted the conversion of loan money from India to be used as the Russian contribution for capital money. On 23rd March 1999, Inter Government "Memorandum on Financing of JV" was signed by Mr. Lambah on behalf of Govt. of India. According to this agreement, it was decided that both India and Russia will jointly contribute 250 million dollars to the project at the ratio of 50.5% and 49.5%.

Over all, rapports were the cause for the establishment of BrahMos. The friendship between Dr. Kalam and Dr. Yefremov believing each other and their team of experts, influence of Mr. Ramanathan with the Indian Finance Ministry and the confidence by NPOM and DRDO on me as a dashing performer to make the JV success revolved around the formation. Whenever I raised a proposal, Dr. Kalam had never rejected it. Similarly, whenever Dr. Kalam forwarded a proposal to Mr. Narasimha Rao, it usually came back on the same day with his approval. It was the confidence Mr. Narasimha Rao was having on Dr. Kalam. A similar rapport existed between me, Ramanathan, the then Financial Advisor (Defence Services) Sivasubramanian and Dr. Kalam. So proposals sailed forward with quick winds in the bureaucratic channels.

On the Russian side, the academic stature of Dr. Yefremov and his legendary role as missile designer helped to move things faster. His name and stature was able to get many things done in Russia. This, I have seen with my own eyes.

On 8th of July 1999, the company BrahMos was officially operational with funding. On the very next day a tripartite agreement was signed between, DRDO, NPOM and BrahMos by Dr. Kalam, Dr. Yefremov and me respectively sharing the responsibilities for the three organisations. This agreement was yet another novel modality where the shareholders themselves became the contractors for the job to be done. By that agreement, BrahMos - a private company, gave contracts to DRDO and NPOM to develop certain systems. DRDO was to develop and provide the guidance components, airframe and launcher systems for flight testing. NPOM was to provide the liquid engine, booster and seeker for flight tests.

We started the JV in a very humble manner. You could have seen the photographs of the founders of today's IT giants like Infosys, Microsoft and Google in their benign beginning days. BrahMos was also like that. As a rule of thumb, all successes have humble beginnings; a giant tree comes out of a small seed. Our beginning was miniscule and insignificant. Even many in DRDO had personally called me up and advised me not to take up BrahMos, but only to concentrate in DRDO. BrahMos was often jokingly called as a difficult venture. But only our grit, perseverance and courage associated with innovative techniques helped us to succeed.

We did not sit idle during this intervening period of more than three years from registration to operationalization of the company. Even before we were able to get funds, we were conducting periodical meetings and basic design works through the staff seconded to the company by the parent organizations. So when we received the funds we were ready with our designs, modalities, schedules and even marketing strategies. The reader must understand this; even before developing the product and testing it, we were planning for marketing strategies. For us, marketing strategies were not confined only to the domestic market (Indian and Russian armed forces) but towards the international market as well. Those two years though dormant from the outset, were really the formative years for the company. Any company or venture needs a thorough planning of all the uncertainties and the contingency plans. Based on this analysis some flexibility had been incorporated in the plans.

8

MOVING AHEAD

Formation of a joint venture company was a new phenomenon for DRDO. Obviously it could not be situated in the premises of DRDO and had to have a secluded place with adequate security. This is because our Russian partners had to come for meetings and few of them had to stay to work with us. As I said earlier, everyday was buzzing with activities. These movements were not supposed to attract the attention of people around because our project was a top secret project. But definitely non-native people will always attract attention. So we rented a building in the area called Vasant Vihar, a residential locality in Delhi with initial ten employees. The locality had many foreign embassies and that gave us a perfect cover for the visiting foreigners. People considered the office just as a sort of an embassy. There was no sign board for the office and even today we don't have any!

Initially, the board of directors had been formed; but just directors only cannot run the day to day activities of an office. So we were to hire employees but how? We gave an advertisement in the newspapers, "A leading MNC needs personnel for the following vacancies, interested may walk in." Thousands of people could have seen this advertisement and 99.99% of them should have thought that "which MNC would advertise without revealing their name in the advertisement." So only a very few who wanted to explore this advertisement came forward for the interview. The venue for the interview was an apartment operated as a DRDO guest house. Many walked away even before the interview. The interview was conducted and those who were selected were intimated through telephone to arrive at a particular address for taking charge. No letters were issued. They were amused for not being informed through formal documentation. After being selected some of them felt that the company was bogus and left even before taking the charge. I and some of the key people addressed them to reveal about the company and about the missile. They were not even able to believe their ears that they were going to take part in India's top secret missile development programme. Those who took charge then, hold some of the prestigious posts in the company today. Everybody was given a specific work with separate instructions. A person's work was not to be known to the other. This was to reduce the flow of information and to plug the information leak. For example, an employee would be asked to draft an important and secret communication without his/her counterpart knowing. Another employee would be sent somewhere and was not available in the office for many days. Others neither asked regarding their absence nor that particular person revealed where he/she had been. There were so many mysterious modus operandi like this.

As in other companies, we also had a front office management system but our front office managers never revealed anything about the company to the callers. The first query from them would be to identify the caller and any unknown calls were not entertained. When calls were made from the office to other offices, even if they are government ones, our identity were never disclosed. The norms were so strict that no government vehicles were used in our day to day operations.

An important identifier for any company is its logo and the standard company communications are made through the company letter head. But to maintain secrecy we did not use any. All our communications had a word BrahMos and that's all. No appointment orders were issued in the letter head for the appointments or any similar purpose. The office that we had rented was a bungalow. We used the servant quarters as the rooms for the staff and the underground cellar served as our library and conference room. We took risk of welding all the windows so that they could not be opened. This was because there were possibilities for theft. In the office there were computers and computers mean data. For us, the data going to anybody else's hand means that the secrecy was over. So we had to take such extreme precautions.

In the laboratory level, the R&D works were simultaneously being carried out both in NPOM and DRDL. The DRDL didn't even named or used the word BrahMos rather a code name PJ-10 was designated. This period continued till our first flight trial. Only after the first trial, the name BrahMos was shared with media for the first time.

GIVING AN IDENTITY

The JV Company has to be given an identity. What are the identities of a company? It has a name and a logo. We needed them. But getting them was a strange experience and that too through strange process.

Naming the Company

There is an interesting story in christening the name of the company and the missile. After the Indian Government approved the joint venture, we, the core group from DRDO and a team from NPOM headed by Dr. Yefremov were on a discussion to coin name of the venture. A thought came on Water being a vital resource for life. It is one of the five elements of nature, gave a wonderful lesson for mankind to learn about vibrancy. It is in day-to-day life helping to preserve the health of an individual. Similarly an organization will be healthy only when it is vibrant. Vibrancy of an organization lies in its thirst to grow through a progressive vision towards the future. The thirst for growth lets the organization to scale summits and venture into deeper oceans. For an organization, growth comes from well defined tasks based on the vision. The fury of water in Brahmaputra, the mightiest river of India bearing the name of Brahma - the creator, is a source of inspiration and vibrancy. Neva is another mighty river in Russia. Both the teams were looking for a combination of the names of the two rivers. NPOM suggested the river Moskva as Russian identity, as NPOM is situated near the bank of river Moskva. We tried various combinations and finally the name came as BRAHMA-MOSKOVA resulting **BrahMos** – combining the fury of Brahmaputra and the grace of Moskva River. I was happily thinking of the coincidence that the name of celestial weapon created by Lord Brahma, known as the Brahmastra, is now given to our missile. All in NPOM were convinced with the name due to the historic background of the two rivers one at the feet of goddess Shakti at Guwahati (Assam) in the eastern India and the other flowing near the seat of power Kremlin. Finally, the Company was named as BrahMos and later renamed as **BrahMos Aerospace**. In the later years, we christened the missile as **BRAHMOS.**

Logo for the Company

We were looking for an appropriately designed logo for the company which should be evolved to reflect the theme and purpose. So we consulted National Institute of Design (NID) Ahmedabad and gave a project for the design of logo. Several options came out of which one was chosen to highlight the following:

- The mighty forces of two rivers and the two great countries joined together to create the power which is equivalent to Shiv Ling.
- Two lines symbolizes the two organizations DRDO and NPOM with their varying capabilities, culture, language, aspirations, coming together to carry on one project representing the speed.

BrahMos

The BrahMos Logo

This logo was approved by the Board of Directors.
This logo is decorated at the atrium of our Corporate Headquarters in a magnificent manner in colour mosaic depicting the Brahmaputra River flowing on the feet of the mighty Kamakhya Devi, Guwahati and Moskva River flowing at the side of St. Basil's Cathedral, Kremlin, Red Square.

Symbolic contribution of Partners – idols of Power and Precision

I narrated about the depiction of the name BrahMos in the atrium in the form of mosaic. The thought came to my mind why can't the prime qualities of our missile be depicted inside the building. The construction of the building itself is separated into two wings. An idea came to my mind that since our product is a Weapon i.e. Missile and the upper portion of side walls at the left wing can be carved with the scenes of the mighty Astras (Weapons) being used in the great war of Mahabharat – An Indian Epic. At the same time, at the centre of the wing, we have constructed a sculpture denoting a scene from Mahabharata, where Arjuna aims at a fish rotating above, by just looking at the reflection of that fish in the water below signifying the precision of the missile which is provided by the Indian guidance system. Similarly, in the right wing of the building, the upper portion of side walls is carved with the scenarios depicting great wars which the Russians had fought. Like the statue of Arjuna, the sculptors of medieval legendary Russian Warriors Ilya Muromets, Dobrynya Nikititch and Alousha Popovitch who were famous for their physical strength have been erected. Their muscle power represents the Power of BRAHMOS derived from the Ramjet Propulsion. Whoever visits the BrahMos Headquarters complex never fails to appreciate the ideas of placing these statues in the premises.

Valour of Russian Warriors
POWER

Arjuna
PRECISION

PART 4

Gaining Momentum

"All one has to do is just look around, find what the world does not have, and then head towards inventing it."

- Thomas Alva Edison

In this part, I will discuss the BRAHMOS missile system and about the force multiplication effect provided by it for our Armed Forces. In a nutshell the missile has rightfully fetched the name as "the World Leader among Cruise Missiles" through its characteristics and qualities. Indian Armed Forces are getting the superior BRAHMOS, which we have achieved through the voyage. But we are not going to stop there but instead develop more and more new weapon systems through our self-found path. The momentum picks up now . . .

9

DESIGNING THE MISSILE

The Gulf war in 1991 gave a clear message that the first level attack on the enemy has to be through cruise missiles with high precision engagement. Our intention to possess an advanced cruise missile system was very clear, but it is necessary to take lead by developing Supersonic Cruise Missile realized in the least possible time and by establishing a brand value. What is that which can help us to leap-frog and minimize the time taken from drawing board to front line deployment, with operational effectiveness of the product and customer delight? This was the question before us.

Through IGMDP we established higher level of capability in Inertial Navigation System (INS), On-Board Computer (OBC), avionics, software and mission management sectors. We required a new engine capable of providing supersonic speed. We had the expertise in Akash project with solid propelled ramjet engine technology. But for the long range cruise missile and for its growth potential, the optimum choice was Liquid fuelled ramjet engine, which is four times energy efficient compared to solid fuelled ramjet. During our consultation with NPO Mashinostroyenia for another project, Dr. Yefremov, the Director General of NPOM was discussing with me on the possible future missions. He asked me to see their ramjet engine developed just before the breakup of Soviet Union. I was amazed to see the engine after learning the specifications I realized that there is no equivalent available in the world. It gave me a feeling of satisfaction that this engine can do miracles. Dr. Yefremov immediately agreed that there is a strong possibility of collaboration in supersonic missile capability.

After assessing the overall characteristics of the engine, our team started looking at the shape of the missile. We realized that the combustion chamber of the engine having large volume which can accommodate the booster. Thus, it will become unified configuration of booster-sustainer combo. This will enable to reduce the length and breadth of the missile, with two-in-one concept. Further discussions revealed that Russian experts had worked out the full configuration of the missile and conducted wind tunnel tests on the model. So a possible configuration was in front of us. NPOM had carried out tests of the prototype and had gone through a few flight trials to establish the propulsion system. Further, it could not be progressed due to shortage of funds in 1990s. It so happened, in Russia, that most of the design bureaus and R&D Centres started losing the best brains as they were moving out to other countries. Having worked with Russian specialists for many years, I felt it was a great loss if specialists leave the country at that critical juncture. A collaborative arrangement with NPOM for the cruise missile would help to retain the best manpower with them. Dr. Yefremov was convinced with this fact.

The initial data on the design of the missile was discussed with the joint design team which appreciated the work carried out by NPOM. Further optimization studies were carried out incorporating Indian subsystems. So with this background, integrating India's guidance avionic systems, the supersonic BRAHMOS took shape. The universal characteristics of the missile for sea, sub-sea and land launch capabilities encouraged us to freeze the design for anti ship application.

Now, we had two options with us. One was to develop an entirely new airframe for the missile. The other was to use the already designed and flight tested airframe.

We had realized five missiles in IGMDP and they were all under the developmental trials. Developing a new airframe based on the existing ramjet modules would have been certainly possible. But to save cost and time, it will be wiser to go for Phase I with Russian design and materials, but fabricated in Indian industry.

Indeed it was a strategic decision. We decided not to reinvent the wheel but to exploit and augment the missile's capability in a step-by-step process. Since, the Russian missile was at a prototype stage, we were sure that its potential could be augmented well with our contribution, especially for land target engagement and warheads.

The next step was to design the guidance system. The reader would be remembering the story of the guidance system developed for Prithvi. It is a state-of-the-art guidance system. The worth of it made itself a fantastic candidate for being the guidance system of the new missile.

Though Prithvi is a high precision ballistic missile, it is an area weapon whereas BRAHMOS is a high precision weapon to destroy targets with a precise hit. So the guidance system of Prithvi needed many improvements. We further augmented the accuracy of the guidance system and also reduced the reaction time, required.

When the discussions came to the point of the launcher systems, fire control systems etc., we decided to develop those systems freshly as state-of-art systems. Today, our land based launcher systems are fully network centric in operation and totally developed in India. There is no cruise missile system operating in the world in the lines of the land attack version of BRAHMOS.

In the hindsight, it was the Indo-Russian partnership that worked well. Each side decided to respect the strength of the counterpart. Neither the Russian partners insisted to use their existing guidance systems nor did the Indian partners insist in developing their own airframe. It could have been possible by both sides but as we operated based on the principle **'strength respects strength'**, we combined the strengths of the two sides. Thus, the propulsion modules became the process stock of Russia and the guidance and onboard electronic modules became the process stock of India, were readily available. The success of the project was that the strengths were combined to form a new product.

Multiple DRDO labs participated in the effort to realize the best design for India. A separate design team was formed to look at the suitable platform system like ship, submarine, road mobile complex and aircraft. The same missile could also be used against different types of targets so the developments of systems like Fire Control System and Launcher for the ship and land were started.

Time proved our decision to be correct, as a missile which was designed to be just an anti-ship missile system has evolved into many fascinating versions capable of attacking the targets that were earlier thought impossible. Thanks to the high accuracy guidance system and high power propulsion system, today we can even hit targets located in a valley.

Not only that, BRAHMOS will soon be getting many new dimensions in the years to come. Can anyone in this world would have imagined about an anti-ship missile used to hit a land based target as well. Nobody would have! Yes, in the forthcoming years, BRAHMOS would be capable of being launched from land platforms, aircraft and ships against sea and land targets in a salvo mode for fast and lethal deployment. There are some more applications in our shelves which we cannot reveal at this stage. The potential of BRAHMOS is huge. It is truly a Brahmastra!

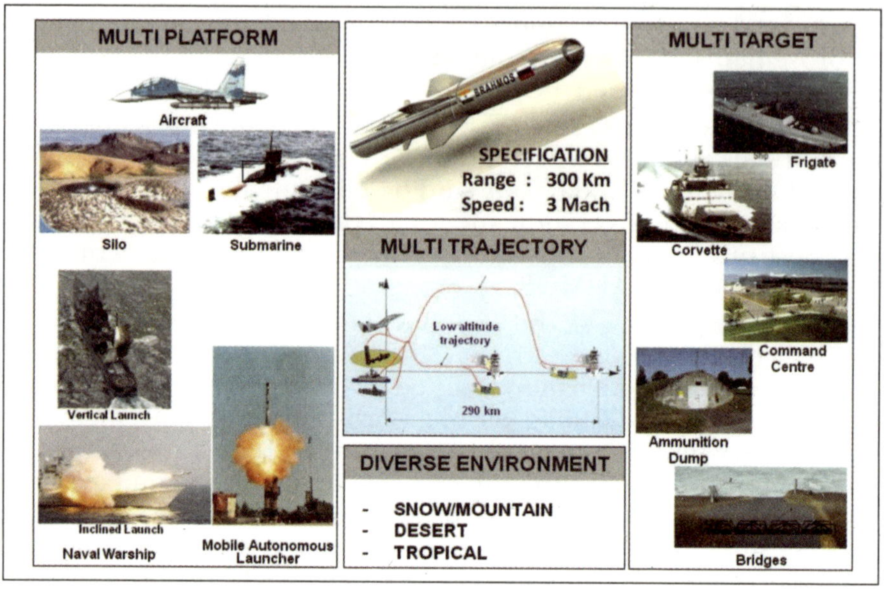

BRAHMOS Unique Universal Missile

This approach resulted in the realization of one of the best designs for the Supersonic Cruise Missile which can be used in multiple platforms against multiple targets in multiple trajectories. What a great opportunity knocked on us to get into the development of fastest cruise missile! We carried out several development trials and saw successive success. Our each flight trial was really enterprising.

10

THE SUPERSONIC BIRD

Alexander Vasilyevich Suvorov (1730-1800 CE) was the fourth Generalissimo of Russian Empire and considered as one of the greatest generals of Russia. Had he been aware of BRAHMOS, then he would have said that 'one second can decide the outcome of the battle!'. As BRAHMOS is the most advanced and most effective cruise missile. In this chapter, I will give an insight about why BRAHMOS is superior and the features of BRAHMOS missile that makes it a system par excellence.

WHY BRAHMOS IS SUPERIOR?—A CAPABILITY ANALYSIS

Cruise missiles are often classified as Anti-Ship Cruise Missile (ASCM) and Land Attack Cruise Missile (LACM). From various estimates it has been concluded that around 75 nations in the world possess cruise missiles. There are around 130 or so cruise missile types operational. An astonishing amount of 75,000 individual ASCMs exist all over the world. Overall there are 19 nations that produce such missiles. Among these 19 nations, about 12 nations produce both LACMs and ASCMs and the remaining nations produce only ASCMs. The most popular missiles in the arms market are the US Harpoon; Russian Uran, Klub; French Exocet, etc.

With around 6000 units having been sold, the US Harpoon is said to be the most sold missile. Then comes the Exocet, which is said to have been sold over more than 3000 units. Among the Russian missiles, the Russian 'Styx' and its Chinese cousin, called the HY-1 'silkworm', have been widely deployed. Russia is currently marketing the '3M-54 Klub' family of missiles which includes the supersonic 3M-54E and the subsonic 3M-54E1 ASCMs as well as the 3M-14E LACM.

Though the statistics are exhaustive but through a keen observation we can see and realize that most of the missiles belong to the subsonic category. Generally aircraft, surface ships, submarines and land based launchers are the 4 type of weapon platforms for launching these missiles. Only a few of them could be launched from all the four.

The "Harpoon (AGM-84)" is an American missile and is said to be the most sold missile. Predominantly it is an anti-ship missile and skims over the water surface with a radar guidance. Some land attack variants also exist. These missiles can be launched from all the four weapon platforms.

"Exocet" is a French missile famous for its anti-ship roles and has tasted blood several times especially during Falkland conflict. The missile can be launched from aircraft, surface ships, and submarines; some land based variants may exist. It is the most sold missile only after to the US Harpoon.

The "Kayak (AS-20)" is a Russian missile whose air launched variant is commonly called as "Uran". This missile is very much similar to the US Harpoon and has active radar guidance. This missile can be launched from all the four weapon platforms and even from helicopters.

The "3M-54" is a Russian missile and generally called as 'Klub' family of missiles. There are five variants in total in this family three for anti-ship role and two for anti-submarine role and one for land attack role. In this family, one of the anti-ship missile is 3M-54E which travels at a subsonic speed and when it reaches the proximity of the ship, the missile fires another rocket that flies at a speed of 2.9 mach.

"Moskit-E" is another Russian ASCM that can be launched from surface vessels platforms. Though this missile is all through a supersonic missile the range is very much limited to only 120 km, not even half the range of BRAHMOS.

Tomahawk is an American cruise missile on which a detailed account has been given already. Tomahawk is generally strategic in nature and can carry nuclear warheads in that role for a fairly very long range. But in our analysis we consider the Tomahawk a Land Attack Missile. These missiles are the largely converted form of the Tomahawk anti-ship missiles that once existed. It flies by maintaining a constant altitude with respect to the ground thus follows the terrain. If there is an elevation in the ground, the missile climbs up and vice versa. But such missiles can be easily detected by the Airborne Warning and Control Systems (AWACS) aircraft that have huge radar. With just a 0.7 mach speed it can be easily intercepted by fighter aircraft. But BRAHMOS, with a speed of 2.9 Mach cannot be intercepted by any current fighter aircraft. Considering the short flight time of the BRAHMOS missile, even before if any aircraft tries to engage the missile, the missile would have accomplished its mission.

AGM-86 is an American strategic subsonic cruise missile launched by the USAF aircraft. This missile could be roughly considered similar to the air launched version Tomahawk.

Listing the features of the famous missiles is alright but how does BRAHMOS is a superior missile?

WHAT MAKES BRAHMOS SUPERIOR?

The supersonic speed is the key for superiority. So far, from the features of the famous missiles it can be noted is that all of them are almost subsonic in speed. But BRAHMOS is a supersonic cruise missile. You may ask, "Just by being supersonic what the advantage is?" I will explain.

The Kinetic Energy (KE) of the body is derived from its speed. Greater the speed, greater would be kinetic energy.

Mathematically the kinetic energy is $KE = (mv^2)/2$

In simple words it can be explained that the KE is half of the value derived by multiplying the mass (m) with the speed (v) squared. Through simple arithmetic we can see that if the speed is tripled then the resultant kinetic energy increases by nine fold. Now BRAHMOS flies at a speed that is three times greater than the other missiles. So it impacts the target with energy roughly nine times greater than the other missiles. What is the benefit of the increased KE?

If the target ship is very huge or the target building is a hardened structure then the increased KE plays a crucial role. For knocking down a huge ship, two or three subsonic missiles may be required. But in such cases just only one BRAHMOS missile will suffice. After learning about the physics behind the impact of the supersonic hit, now it is time for economics.

A single missile does the job of two or three missiles for which the user pays for just one missile and enjoys the effect of three missiles. Not only the down payment is reduced, but the life cycle cost also reduces.

The user has to just keep one missile ready instead of three missiles. In case of cramped space in warships he saves the space of two missiles.

There are many more factors apart from the supersonic speed that make BRAHMOS missile superior. Let us see some.

STRIKING FEATURES OF THE BRAHMOS MISSILE

There are lot of superior features in the missile and associated systems, but for the sake of this book I am listing only the ones that can be revealed. The other features have to remain secret for the sake of security of the nation. By these qualities the missile claims its rightful place as an "Ultimate Force Multiplier for Armed Forces". I shall explain each characteristic for the readers to understand the potential of the missile.

Lethality through Supersonic Speed

BRAHMOS is a supersonic cruise missile. A need may arise to fly BRAHMOS at a very low altitude in order to avoid detection by enemy radar. This type of trajectory is called 'flying on the nape of the earth'. As the earth is spherical, any airborne vehicle flying just like skimming over the ground cannot be detected from long distances through radars. The possibility of detection of the missile is near its target (e.g., ship). In case of ordinary missiles, the enemy gets a considerable amount of time to activate his defences. In case of BRAHMOS, as the speed is greater, the missile gives no time for the enemy to detect the incoming missile and act accordingly. As the speed is three times higher, the enemy's reaction time is reduced by one third.

At the same time, the supersonic speed gives enormous kinetic energy to the missile due to which when it hits the target, it hits with a lethal force. In one of the flight trials, the target was a decommissioned ship. The missile penetrated the entire ship and went away and such was the effect; it was a sight to be seen. The damage caused was entirely due to the missile's speed; if you remember the relationship between the speed and kinetic energy you can understand the effect.

Multi Target Capability

A single missile can hit a band of targets ranging from ships in the high seas to buildings deep inside the enemy land. The missile can hit vessels of various sizes from small corvettes to large aircraft carriers. It can hit a variety of targets like factories, buildings, fuel depots and so on. The user need not stock different missiles for different targets.

Precision Hitting Capability

A missile can be faster and intelligent but in the war scenario what matters is the accuracy of the hit. BRAHMOS missile has the highest accuracy rate among various missiles. Hitting a decommissioned ship at the bull's eye speaks about the precision hit capability of BRAHMOS.

Multi Trajectory Capability

The missile has been designed such that it can follow various trajectories. It can also fly at various altitudes. Generally the missile can follow any of the following four trajectories:

1. High-Low trajectory
2. Low-Low trajectory
3. Waypoint trajectory
4. Steep Dive trajectory

In High-Low trajectory, the missile climbs to a very high altitude. After maintaining the same course for a certain distance, it then dives sharply and follows the nap of the terrain. This trajectory is chiefly followed for anti-ship role as the missile approaches the target suddenly from the horizon and the enemy will be left with no solution except to say his final prayers.

Low-Low trajectory is followed when the target is at a closer range. The missile will fly in the nape of the earth altitude all through the entire flight. When the missile suddenly appears at the horizon, as I said, the enemy has left with no options.

Waypoint trajectory is followed when the enemy has to be deceived. Through this trajectory the missile can be allowed to fly to the target through a different direction instead of the straight line. The missile can be directed to fly such that it can circumvent an enemy air defence position that would be expecting the missile. Though presently the cruise missile defence systems are not capable enough to intercept BRAHMOS but they might be developed in the future. Even at that time, those systems can be cheated by this manoeuvre. Assume that an enemy's sailing vessel is to be attacked. A salvo of four missiles can be launched with each missile to arrive at the vessel from all four directions. Each missile would fly at varied altitude so that the target vessel's defences will be overloaded with threats emanating from different directions in difference profiles. In case where three missiles having being intercepted, the fourth missile can still destroy the vessel.

In the steep dive trajectory, the missile after reaching the target area dives vertically and hits the target over its head. This mode would be highly useful in hitting targets situated in mountains and partially buried command centres. No other missile in the world flies in a trajectory like this.

Multi-platform Capability

The missile is a universal since it can be integrated with various weapon platforms like land based launcher systems, surface vessels, aircraft and submarines. Simply saying the missile can be fired from the land, air, sea, and from subsea platforms against targets on both land and sea thus covering all the layers of theatre. There are so many cruise missiles available in the market having a capability to be fired only from anyone of the platforms. But in the case of BRAHMOS since the missile is universal, it has so many advantages in the war and can be launched from anywhere. During a war, the enemy cannot predict from which platform the missile will be fired. Also it can be used by all the wings of our armed forces. Combining the multi trajectory capability with the multiple platform capability the missile can strike the enemy from various directions and can bring him to his knees.

Predominantly in anti ship missions, the missile has a very good range to hit a ship travelling in the deep seas. Such an advantage makes the missile a terrific weapon of choice for the coastal defence role. Assume that our vessels are on patrol in the deep seas some 250 km away from the coast. When a fleet of enemy vessels is detected some 500 km away from the coast, the entire enemy fleet can be detected and destroyed before they can launch any meaningful attack through their aircraft or missiles. So the enemy is destroyed before he can reach any meaningful distance to launch a strike. Similarly our missiles can hit a coastal target by staying some 250 km well beyond in the international waters.

In the 1971 war with Pakistan, our small coastal defence missile boats were taken up to the entrance of Karachi harbour to hit the enemy vessels and their installations. Those boats did not have the range and endurance to sail for longer distance. Hence, they were towed to a very long distance off the coast of Bombay and then mission got started. Also, the missiles that were fired had a range of lesser than 80 km. But with BRAHMOS there is no need for such hard grinding of our forces. Simply the naval vessels can stay in a standoff range; fire the BRAHMOS missiles against the enemy targets, and then return. Similarly the Indian Army has inducted the BRAHMOS missiles that can be fired from a road mobile vehicle. When a group of such vehicles are concentrated and when they start firing the missiles, the enemy will see a volley of missiles. Just by staying well inside our territory our army can launch the missiles and may even end the war before it begins.

Work is underway to integrate the missiles with the Sukhoi-30 MKI aircraft of the IAF. When integrated with these aircraft, the missiles can be fired over a greater range with impunity, as these aircraft have enormous range. This makes the aircraft and the missile, a deadly combination.

Similarly, the missile can be launched from submarines, which can sneak near the enemy shore and then fire the missile. Also the submarines equipped with BRAHMOS can perform the anti-ship role effectively in the deep seas stealthily.

Multi-mission Capability

BRAHMOS can perform various missions and it can fill the gap between the purely tactical missile and a strategic missile. In air war there are two types of bombing, one is called as strategic bombing and the other is called as tactical bombing. To put it in a nutshell, tactical bombing means, bombing the weapon against platforms like destroying ships, troops and tank columns. Strategic bombing means bombing the war waging capability like destroying factories, oil refineries, dams, etc. BRAHMOS can be used to destroy both tactical and strategic targets since the missile can hit enemy vessels in the deep seas and also various strategic targets as listed before.

Salvo Launch Capability

The missile is independent in nature, for example two missiles, each targetting different targets, can be launched from a same launching platform in quick succession. The two missiles in action do not interfere with each other and this feature enables multiple missiles to be launched either against a single target or against various targets. Most importantly these two missiles can be launched in an interval of just a few seconds. A vessel like "INS Ranvir" has eight missiles and all eight can be launched against eight different targets that too in quick succession. Similarly in land based systems such capability is present. Merely this capability is the key to win future wars. Imagine a squadron of vessels or land based launcher systems launching a volley of missiles to cripple the enemy even before he starts the war.

The Most Cost Effective Weapon System

When compared to the other missiles, BRAHMOS is much more economical in terms of capabilities and power, the cost invested in the missile is nothing. In a war, to destroy an important enemy installation like long range radar, a single BRAHMOS missile can be used to destroy that installation instead of sending a formation of aircraft. This reduces the risk for aircraft and pilots while operating inside enemy territory. At the same time, that formation can be used elsewhere thus an extra formation is available for combat duties. This is called as force multiplying; a single missile provides the effect of a formation of aircraft.

Being equivalent to three other missiles and also considering the multi-role, multi-target and multi-trajectory capability BRAHMOS will be highly economical. The comparison of capabilities of BRAHMOS to that of a typical subsonic cruise missile is given in the following table:

	SUBSONIC	BRAHMOS
Speed	0.8 Mach	2.8 Mach
Time to hit the target	1 unit	1/3rd (Faster engagement)
Kinetic Energy	1 unit	9 times. (High Destructive Power)
Target Dispersion (Moving targets)	1 unit	1/3rd (Probability of hit is high)
Reaction Time	1 unit	1/3rd (Pierces the Defence)
Universality	Nil	Same system for sea & land targets

Longer Shelf-life and Easy Maintenance

BRAHMOS missile has a longer shelf life as compared to any other missile. During the production time the missile is assembled and inserted in a canister. Like a scabbard protecting a sword, a canister (cylinder) protects the missile. This canister seals the missile in air tight condition. Thus, the environmental conditions do not cause any effect to the missile. Particularly, the missile is well protected against oceanic corrosion while present in warships. Similarly, the land attack variants are also protected from the harsh weathers ranging from desert to the icy heights of Himalayas. Apart from being protected, the missile is very easy to maintain. In general, a sophisticated weapon system or platform need not be simple for maintenance. Usually lot of man-hours are spent to keep the system in 'fighting fitness.' In the case of a famous American missile, they have to be tested once in a quarter whereas BRAHMOS can be tested once in three years. The missile has a longer shelf life because of the protection offered by the canister which allows the missile to be transferred anywhere. When the missile is in canister there will be no contacts with the outside world. Once in three years the status of the missile is checked through umbilical connectors.

Low Radar Cross Section

Radar Cross Section (RCS) is a parameter that is used to indicate the reflectivity of an object to the radar. Generally greater the surface area of a body, greater is the RCS thus making it visible in radar precisely. A civilian transport aircraft reflects lot of radar energy back to the radar than a fighter aircraft and is highly visible in the radar. A cruise missile produces a RCS lesser than the fighter aircraft. Cruise missiles with deployable wings reflect considerable amount of RCS. Since BRAHMOS does not have deployable wings and is very compact, the missile does not produce much RCS. Lower RCS reduces the detection capability of the enemy radar. Thus, without extra efforts, BRAHMOS is inhcrently stealthy.

Intelligent Behaviour

The missile has a number of intelligent behaviours and one of the important behaviour is the prudent usage of the radar seeker. Assume that in pitch darkness you are supposed to find an enemy with a flashlight and he is also supposed to do the same. You can find the enemy using the flashlight and also without flashlight. How? When your light is switched off, then there is only one light beam, which will be of the enemy. From that light beam you can find the location of enemy's flashlight. The radar can be substituted for the flashlight as the radar also behaves exactly like a flashlight.

Detecting an enemy's presence by detecting his radar's emission is called as 'Electronic Support Measures' (ESM). Once, the enemy's position is detected then the job becomes easy to confuse or counter him. So when we have radar receivers, we can receive the enemy's radar signals and without using our radar actively, we can find out that the enemy is coming. Often a condition called as Emission Control (EMCON) is imposed on the forces that they have to stop emitting the radar and communication signals. This was a lesson learnt after the Second World War. The Japanese attacked the Americans in the Pearl Harbour where all their aircraft originated from Japanese aircraft carriers. The pilots had an instruction that after a certain distance, they have to tune their radio receivers to the Pearl Harbour radio station's frequency. The unsuspecting Americans did not impose any restrictions on the broadcasting radio station. These signals are directional and we know this since radios in our homes receive well in one direction. So using these signals, the Japanese reached the Pearl Harbour without much navigational hiccups and inflicted a heavy damage. From that incident, predominantly, all the militaries learnt to reduce their signal emissions. But they compensate this reduction by listening to the enemy signals. And when they detect any emission then they initiate a counter measure called as Electronic Counter Measure (ECM), where the enemy radar signals are confused and jammed.

BRAHMOS uses its energy efficient radar seeker in a diligent pattern. The radar signals from BRAHMOS do not attract the attention of enemy ESM systems.

Effective ECCM Features

ECCM is a terminology that stands for Electronic Counter Counter Measures. ECCM means countering ECM by making the electronic systems more robust and jam proof. BRAHMOS has sophisticated ECCM systems, so that ECM of the enemy will not pronounce any effect on the missile. Considering the supersonic speed of the missile, the ECM of the enemy cannot create any damage since the reaction time for him is very less.

Universal Launch Configuration

BRAHMOS can be launched from an inclined position, vertical position from surface ships and also can be launched horizontally from an aircraft. We have fitted the missile in the inclined configuration in INS Rajput. If the user does not require much complicated alterations, this configuration is the ideal one. But this comes at the cost of some combat potential. The maximum combat potential lies in vertically launched missile where it has all the freedom to align itself with respect to the target. Only BRAHMOS is capable of doing so. This configuration can fit in all sorts of surface vessels and submarines. In the air launched version, the missile would be dropped horizontally from the aircraft to allow a safety clearance from the aircraft. After a particular distance, the missile will fly towards its intended target. The capability to be launched from any angle makes the missile more robust for the demanding combat environments.

SECTIONS OF THE MISSILE

The missile has five major sections namely; Propulsion System, Nose Cap, Target Seeker, Airframe, Guidance & Control System and Warhead. Each system in BRAHMOS is uniquely designed and has multipurpose functioning. I will narrate each system in brief in succeeding paragraphs.

Propulsion System

The propulsion system of the BRAHMOS missile is the most unique among any flying machine! This system can be called as composite propulsion system as it contains a solid rocket booster and a ramjet sustainer. You know the ramjet engine can function only after attaining supersonic speed. So, to propel the missile to a required speed, a solid propellant fuelled rocket booster is used. Solid propulsion systems are quick and do not need any tedious maintenance procedures. In many contemporary cruise missiles, Booster is attached to the exterior. In BRAHMOS, the booster is housed inside the ramjet engine. But it causes hindrance for the ramjet engine to work as the additional weight of the booster remains of no use after its work is over. So the booster is jettisoned after the required speed is achieved. The space being occupied by the booster becomes the combustion chamber of the ramjet engine. This way the missile becomes highly compact which helps it to be stored inside canisters.

The missile is ejected from the canister through a gas generator system, colloquially called as gas bottles, situated inside the canister. The generated gas remains at a high pressure pushes the missile out of the canister. Once the missile leaves the canister, another system called as Low Thrust Motors (LTM) fires and takes the missile to a minimum altitude after which the main rocket booster takes over safely. Do you know which fuel is used for the ramjet sustainer engine? It is Aviation Kerosene, a highly purified form of kerosene.

Nose Cap

Nose cap is used to seal the missile in the canister. It also protects the missile during underwater launches, i.e., when launched from a submarine. Apart from the protection, it has a combat utility. During vertical launches, it brings the missile to the proper course. Simply saying, it tilts the missile horizontally and orients the missile in the direction of the target, before the solid rocket booster fires. Say if this feature is not available and if the target is in another direction to the missile, the missile has to make a 'U' turn to reach the target. This would result in wastage of fuel thus reducing the range and also the precious time. To avoid this, the nose cap twists the missile towards the target after it is being ejected from the canister. Now the target would be right ahead of the missile and again the nose cap works to tilt the missile to a horizontal position. But after these manoeuvres, the nose cap itself becomes an unwanted load for the missile and hence gets ejected. BRAHMOS is the only missile in the world that functions in intelligent manner.

Target Seeker

The target seeker is basically a radar seeker. The seeker scans the area in front to find the target. An improved seeker has been developed for the Block II version of the missile. These seekers will be used for the land attack version of the missile for the army.

Airframe, Guidance and Control

The airframe of the missile is a perfect cylinder with small wings and fins. This design is the predominant feature in reducing the drag at supersonic speeds. Wings and fins are kept folded around the missile body inside the canister and when the missile leaves the canister they open up and get deployed. The wings if you see are very small because even a small wing can provide sufficient lift at supersonic speeds. This is the reason why fighter aircraft have delta shaped small wings whereas airliners have huge wings. The smaller wing profile creates a very negligible RCS, as I said earlier.

The wings are fixed and the fins move to turn the missile during the flight. There are four actuators, located to actuate the fins. These actuators are commanded by the guidance system. The air intake for the ramjet engine is present in the nose of the missile. This also reduces the RCS whereas belly air intake usually leads to causing more RCS. Where will the liquid fuel be stored for the sustainer engine? The fuel tank is an annular fuel tank shaped in such a way that the body of the missile itself acts as the wall for the fuel tanks. The missile is very small in size when compared with the mechanical and electronic complications that are involved. The entire area is used for accommodating various subsystems and indeed a highly complex arrangement. Explaining further will make no sense as it would be purely technical.

The guidance and control system is based on an INS which works in conjunction with an On Board Computer (OBC). During the flight, the INS detects the deviation from the intended flight path and feeds the OBC that in turn directs suitable aerodynamic features to restore intended flight path. The target's coordinates are fed to OBC via a Fire Control System (FCS) present in the launcher before launching the missile. The missile then flies to the target area by deriving inputs from the INS and Seeker to derive its trajectory.

Warhead

The warhead of the missile can store 200 kg of High Explosive (HE) material. In one of the flight trials, we fired the missile with a live warhead. The missile flew and hit the target - a decommissioned naval vessel and it sank within 6 minutes. When we went to inspect the target after the hit, there was no target and there were not even much debris. Similarly, when the same warhead is used against a land target like a building, only debris would be left out. Such effect is possible because the lethality of warhead is augmented by the supersonic speed of the missile to create a terrible damage.

Not only the missile is sophisticated but its launcher systems and fire control systems are also equally sophisticated. I shall explain about them now. There is “Ship Based Weapon Complex” for the Navy and “Road Mobile Weapon Complex” predominantly for the Indian Army. The same version that the army uses would be used by the Indian Air Force.

11

MULTI PLATFORMS: UNIQUENESS

The strength of BRAHMOS missile is further augmented through the platform system complexes. BRAHMOS can be launched from various weapon platforms. We call the launch platforms as platform systems complex. There are ship based, road mobile, aircraft based, submarine based and silo based weapon complexes. In this chapter, I will explain how BRAHMOS missile will be operated as a weapon system through the platform systems complex. We will see them one by one.

SHIP BASED WEAPON COMPLEX

The ship based weapon complex consists of launcher system and a Fire Control System was first installed in an R-class destroyer namely INS Rajput. Indian Navy is operating five such vessels; these are INS Rajput, INS Ranvir, INS Ranvijay, INS Rana, and INS Ranjit. The destroyer was originally built in USSR and called as Kashin class destroyers; Rajput class is its Indian name. These five vessels form the backbone of the striking power of Indian Navy.

The INS Rajput was originally equipped with the P-20 missile in an inclined configuration. This P-20 missile was replaced by BRAHMOS. Then as a next development we installed BRAHMOS in the vertical configuration in the vessel INS Ranvir. The important process is integrating the missile complex electronically, physically and mechanically with the vessel. We will see the individual systems one by one.

Fire Control System (FCS)

As the name indicates, it controls the firing of the missile and it is in this system the launch control switch is situated. The prime function is route planning where it is decided how the missile should fly and reach the target. To reach a target, the missile can be made to fly directly in a straight line or in a curved path. Apart from this, FCS maintains a direct contact with the missile and reports about the health of the missile. When we went for INS Ranvir installation we improved the previous FCS with the Mark-II FCS.

In a warship the captain's chair is situated at the tallest point of the vessel and captain's cabin is called as 'Bridge.' It is from the bridge the sailing and combat operations were directed in the past. This was because all the naval engagements during the olden days were carried out in the vicinity limits. So the captain had to be in the top of the vessel to see up to horizon and give his commands. But after electronics started dominating the war field, the distance of engagements increased.

The contemporary ships have various sensors as the eyes and ears of the vessel, and have missiles as the main firepower. All the inputs from the sensors and weapons status systems are pooled into one station called as 'Combat Direction Centre (CDC).' The captain can now conduct the battle from the CDC, he can fire his missiles, his guns, etc. from CDC. The brain of the CDC is called as Combat Management System (CMS). This CMS will be connected with the FCS of many weapon systems present in the vessel. All the sensors like radar and sonar are connected with the CMS from which the FCS takes its inputs. Man Machine Interface (MMI) is the buzzword in the contemporary day weapon systems because easier the operation, greater is the efficiency.

Coming back to the FCS of BRAHMOS, when we designed this FCS, we kept in mind the MMI as the top priority.

FCS has a number of Graphical User Interface (GUI) systems for effective MMI. FCS Carries out Self-checks, Link checks with external systems and checks missile's health and displays all these parameters. It receives target data from CMS or accepts the manual entry by the operator. It also facilitates the operator by providing the combat and route planning for the missile when the target coordinates are provided. In a nutshell, FCS is an interface between the missile and the operator.

Inclined Launcher System

BRAHMOS has an additional capability to be launched from the inclined launcher. In one of the initial test firings, the missile was test fired from a stationary inclined launcher. The agenda of the flight trial was to prove that the missile can be launched from an inclined launcher. As the viability was proved, Indian Navy gave the INS Rajput to install the inclined launcher. There were four P-20 missile launchers in the vessel; one was removed and BRAHMOS launcher was fitted in the bow (front side) of the vessel.

When the target is in the front of the vessel then there is no operational problem and the missile can be fired directly towards the target. But in a war, we cannot plan as per prediction. There might be an enemy vessel trailing our vessel to attack. In such a case, the missile may be launched and programmed to take a circular course towards the enemy. There are some limitations for the missile to take a turn because the greater speed limits some sharp turns. So we planned a launcher that can launch the missile vertically from the ship. We developed the launcher and gave the name Universal Vertical Launcher Module (UVLM).

Universal Vertical Launcher Module (UVLM)

The UVLM was designed by the BRAHMOS team and its design has been patented. The launcher got its name because it can be directly used in any kind of vessel like submarines, destroyers and frigates. This UVLM is the outcome of our idea of designing only one launcher for any ship. We wanted one launcher as a system to be fitted on any type of surface vessel or submarine. The greatest advantage of the system is that it can store more missiles than that of inclined launchers.

Even one or more UVLM can be installed in one vessel. Imagine if 2 UVLMs are used then the single vessel will comprise 16 such missiles which mean 16 different targets can be attacked in quick succession. Imagine a group of 3 Ships each with 16 missiles. Then a total 48 missiles are readily available to hit 48 targets with an unspecified number of missiles in storage.

The UVLM has top opening hatches. When the missiles are to be launched the hatches are opened by the FCS. The firing of one missile does not affect other missiles. The unique design of UVLM spreads out the shock resultant from the missile exit evenly, and also ensures that the shock does not affect the vessel's structure.

Technological Hurdles We Faced

The general principle in ship building is that around 20% of the space is reserved for installing systems that would be developed in the future. Because a ship can serve a navy for around 40 years, but during these 40 years there would be sea changes in principle of war and the weapon technology.

The challenge for us was integrating BRAHMOS with an old vessel; it was not that easy. If you see the dates it will be clearly visible. INS Rajput was commissioned in 1980s but the missile got integrated in the year 2003. The electronic systems belong to the 1970s era, and the vessel should have proven systems. It was a great challenge for us because there were lot of problems in integrating the missile with the vessel. But undeterred by the challenges, the weapon complex was installed in INS Rajput in record time.

After INS Rajput got BRAHMOS edge, the time for the INS Ranvir's maintenance and repair came. As a follow up, this time Indian Navy asked us to install BRAHMOS missile in the vertical launch configuration. Foreseeing this development, we were ready with our UVLM by this time. When we were installing the inclined launcher in INS Rajput we were ready with the design of the UVLM. By the time, INS Ranvir came for maintenance we had validated the design and realized it in record time. Such activities are highly time bound and without proper planning only chaos occur.

These R-class vessels did not have any vertical launchers. In the case of INS Rajput there was a launcher for another missile which was replaced rather easily. Now for Ranvir we had to alter the vessel by cutting some of the portions in the upper deck, remove them to accommodate the UVLM. A tedious process of the degutting and integration was carried out in the docks of Vizag, a coastal city in the eastern seaboard. We were having a feel that fitting BRAHMOS in INS Rajput was tedious. But after encountering the problems in Ranvir we felt that Rajput project was cake walk. Finally, now INS Ranvir was also given the BRAHMOS edge and a flight trial was conducted successfully.

After the successful installation of BRAHMOS weapon complex in R-class ships, the Indian Navy also asked us to install the weapon complex in the vessels like; "P-15A", the new Kolkota class destroyers; "P-15B"; and the "Talwar" class stealth frigates. The vessels P-15A and P-15B would be equipped with16 missiles whereas Talwar class gets 8 missiles.

At this point, it will be prudent to elaborate the potency of the ship based weapon complex. In the figure you can see a sketch of a naval attack scenario in salvo mode. Here, an early warning helicopter belonging to the squadron is patrolling well ahead of the squadron of vessels. This helicopter with its high power radar detects the presence of enemy vessels and relays the directional coordinates of the targets to the squadron. After proper authorization, the vessels will fire BRAHMOS missiles according to the requirement.

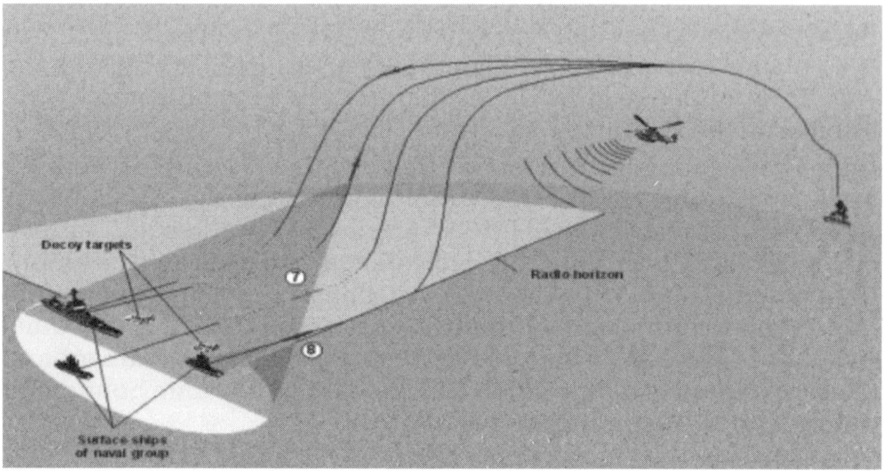

A typical engagement sequence

In the figure you can also see some decoy targets, Decoy targets are created by chaff clouds. Chaffs are nothing but thin foil of aluminium like metals that can reflect the radar signals. Generally, radar reflects signal, it means there is possibility of a target. So chaff clouds can deceive a radar by posing itself as target. But BRAHMOS has adequate intelligence to discriminate the actual target from the false ones and can accurately home on to the target.

A group of some small vessels with an early warning helicopter would be a deadly combination for the enemy to defeat. But if ocean surveillance satellites are operated to scan the area continuously then they can also send target locations to the vessels directly. Such a surveillance platform will be the order of future wars.

ROAD MOBILE WEAPON COMPLEX

There are two variants of land based weapon complex; one is the land attack variant and the other is a coastal defence variant. Both the variants are almost similar in nature. The prime constituents of the land based weapon complex are a Mobile Command Post (MCP) and Mobile Autonomous Launchers (MAL). One MCP commands four MALs which is called as one Battery - a basic unit. The MCP in turn gets its command from a Command, Communication and Control (C3) centre. The MAL and the MCP are communication intensive systems having a robust and redundant wireless voice and data connectivity. As an important feature, the communication is thoroughly encrypted to prevent the enemy from intercepting the communications. Some techniques are followed to prevent any jamming attempt by the enemy. In the battery there is workshop vehicle used for conducting repairs, a Mobile Replenishment Vehicle (MRV) used to load the MAL with missiles and also Light Combat Support Vehicles (LCSV). We will see one by one from the top to the bottom.

Mobile Command Post (MCP)

You would have got the glimpses of our weapon systems development policy from the IGMDP section; "the system that is developed should be futuristic." In that line BRAHMOS has not only been developed for today's war scenario but also for the future war! The future wars are going to be based on Network Centric Warfare (NCW) and the road mobile weapon complex is a starter for that. MCP fits as one of the primary node in that NCW hierarchy.

The MCP takes the commands from the higher echelons regarding the targets designated for that particular battery. The MCP designates the targets for each MAL, gives the target coordinates to the individual MALs, fixes the number of the missiles to be launched, and the time of launch. The MCP monitors the status of all MALs. In case if any one of the MAL has developed a glitch then another MAL from the battery is assigned that task. Depending upon the terrain and situation, the MCPs are delegated the tasks which are sent turn-wise as commands to individual MALs.

Similarly in a coastal battery, the MALs are given a set of targets to destroy and also instructed whether to launch in salvo mode or not. As I have said earlier war is dynamic in nature where targets and target priorities can change at any time. Any change in target assignments is also catered for. The data of all the events in the MALs before and after launch are recorded based on which the After Action Report (AAR) is generated.

Another important function is the analysis of Terrain which is done from the cartographic data to select a suitable deployment site for the MALs. The area through which the missile has to fly is also analyzed. Say from a MAL the target may be at a short flight distance but if there is a hill then the missile has to circumvent the hill. Such issues are addressed by the MCP. As a true command centre, the MCP has a clear tactical picture of its Area Of Responsibility (AOR) that shows the MAL deployments and threat scenario.

Mobile Autonomous Launcher (MAL)

MAL is a missile launcher system capable of carrying and launching three missiles. But why there is the word autonomous? Because autonomy is the key feature in this context. You might have seen Agni and Prithvi missiles in the Republic Day parade. In parade, we show only the missile launcher and not the entire system. There are numerous vehicles for various requirements to support the launcher.

We wanted to break this dependency by making the MAL a self contained vehicle. MAL has a launching station, generator systems and many more systems. This is the reason why we gave the name, "Mobile Autonomous Launcher." Though there are some support vehicles in the battery but MAL is a highly independent system. Also, there could be a technical hiccup with the MCP or in an unfortunate moment the MCP may also get destroyed by enemy action. There is a military saying "Never ever underestimate your enemy" and "Anything can happen in the war". So keeping these dangers in mind the MALs have been designed in such a fashion that in the event of a loss of MCP, any one of the MAL can take over the role of MCP.

The FCS is the predominant unit of the MAL similar to the FCS in a ship based weapon complex. In the MAL the missiles are kept in the horizontal position during storage and transportation. They are articulated to a vertical position at the time of launching.

MAL is capable of either launching the missiles in a single or salvo mode. Now as you know, one missile battery contains four MALs which make a total of 12 missiles, each can be fired in rapid succession. So a missile battery has a firepower packed to bring down 12 targets within minutes.

AIRCRAFT BASED WEAPON COMPLEX

Currently work is underway to integrate the missile with the Su-30 MKI aircraft for the Indian Air Force. The Su-30 aircraft has a phenomenal range. When armed with BRAHMOS they will fetch a name as a deadly pair. Initially we gave proposals to Indian Navy for integrating the missiles with Tu-142 and Il-38 maritime reconnaissance aircraft of Russian origin. Tu–142 is the modified version of the Tu-95 intercontinental strategic bomber with an awesome range. The integration of BRAHMOS missile would have made it the king of seas and could have kept the enemy naval assets under eternal threat. But Indian Navy stated that those aircraft had lived their lives and were in the verge of retirement.

So the only option at hand is the integration with Su-30MKI. The IAF has assigned 2 aircraft for modification and conducting of BRAHMOS trials. The aircraft modifications are being carried out at Hindustan Aeronautics Limited (HAL).

Simultaneously, we have undertaken the design of the airborne launcher. The launcher is as important as the aircraft modification. The aircraft will undergo aerodynamic validation tests for the launcher by carrying it under its belly. The launcher has been designed & developed by BrahMos Aerospace Thiruvananthapuram. Soon BRAHMOS will get integrated with the Su-30MKI aircraft to establish its reign.

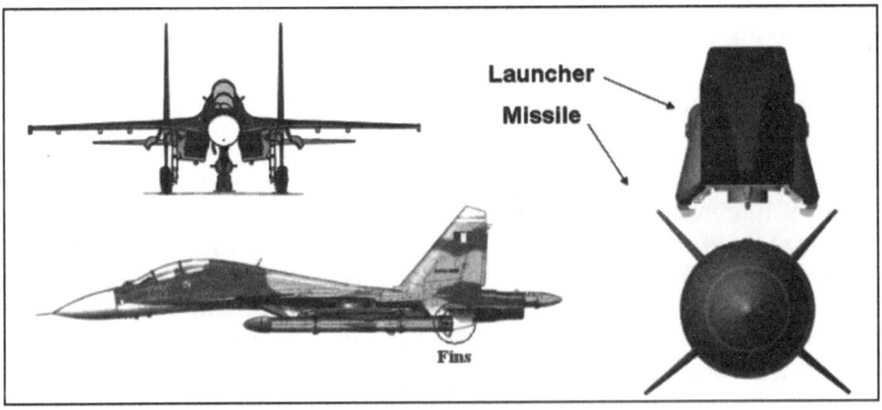

Fitment of the missile in the Su-30 MKI aircraft

BRAHMOS – A is the name we have given for the Air launched version of the Missile because it is slightly different from the present version. The missile has undergone some modifications for weight reduction. You may remember that the propulsion system is composed of a solid propellant booster and the ramjet sustainer. The booster is supposed to accelerate the missile to a supersonic speed from zero speed. The missile already being carried by a fighter aircraft at high altitude means that it is already being exposed to sufficient fast air flow. Whilst the booster does the same job, for this version, it need not be heavy since its extra weight would consume extra fuel from the aircraft thereby reducing its endurance. So we are reducing the size of the booster for accelerating the missile to a sufficient speed.

The BRAHMOS-A missile's launcher holds the missile. At the launch command from the pilot, the missile is dropped horizontally. The missile undergoes a freefall for a certain height, then it stabilizes itself and only then the booster is fired. During the free fall and the firing of the booster, the missile has to be in a stable orientation. For adding stability we have added small tail fins to the missile.

Now you might be wondering why there is no canister present for the air version. It is because along with the weight of the canister the missile cannot be effectively carried by the aircraft. But if there is no canister then what is the use of nose cap? The missile will be carried by the aircraft in a slightly rolled orientation leading the wings of the missile representing a cruciform ('X' shape). Now, when the missile is dropped from the aircraft it falls to its designated altitude and then the nose cap will fire to roll the missile to its normal position. Then the main booster will fire. After this sequence of events, everything is similar to the present version. It is easy to write but extremely hard to realize. It is not that it can't be realized but it has taken sometime and soon the missile is going to be integrated with the aircraft.

Peripheral Control Device

The aircraft has its own computer called as flight computer as the centre piece of the avionics. The flight computer gets inputs from all sensors like radar etc. and communicates with the pilot through displays. From the movement of various pilot controls, like throttle stick etc., it understands the intentions of the pilot regarding the intended flight path. Accordingly, it actuates various aerodynamic features. Actually it is only the flight computer that flies the aircraft, but it flies according to the pilot's instructions, which he gives through his flight controls. The missile has its own OBC that directs the missile to the target. Now this computer needs the location of target which is supplied by an Peripheral Control Device (PCD) which is FCS for air platform, present in the launcher itself. This PCD communicates with the missile's OBC and the aircraft's avionics.

The pilot in a typical anti-shipping mission will turn the aircraft's radar in the sea search mode and when he finds his target, he selects it. The coordinates of the target are fed into the PCD. PCD feeds the target coordinates in to the missile's OBC. When the pilot squeezes the launch trigger, the missile gets released from the launcher and starts freefalling. The missile stabilizes itself to fire the booster. Then the routine programs in the missile's OBC are executed. In a similar fashion land attack missions are also carried out. The following sketch depicts the usage BRAHMOS-A missile used in the anti-ship role.

The aircraft can either get the targeting data from the satellites or from the AWACS aircraft. The aircraft with their own radar can also find their targets and engage them. The aircraft can easily be detected by the enemy since it would be flying above the radio horizon and the enemy radars will face no problem in detecting them. Will it not be a disadvantage? The answer is NO!

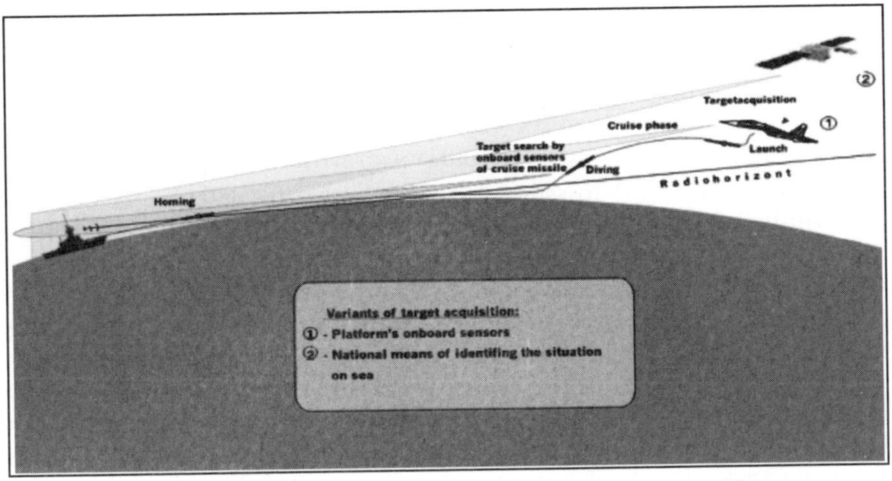

Typical engagement of an enemy vessels using air launched version

The aircraft will fly up to a standoff distance from the targets, launch their BRAHMOS-A missiles and then safely withdraw. The missiles will in turn fly the remaining distance below the radio horizon. As usual, even if they detect, you know it is next to impossible to intercept BRAHMOS. Even the ships will have face difficulty in detecting BRAHMOS even if they are equipped with powerful radars.

SUBMARINE BASED WEAPON COMPLEX

Submarine is a complex and stealthy machine. Present day nuclear submarines can travel at tremendous speeds and remain submerged for longer durations. When armed with missiles they become a deadly hidden predator lurking to take its prey. Today, every nation is trying to get nuclear submarine and install their ICBMs in them. Doing this gives the assurance that the adversary would not launch his nuclear missiles to prevent the retaliation from the opponent.

At the same time, submarine is also a dangerous place to work with many operational hazards which may take enormous toll on the psychological health of its crew. I had an opportunity to release one of my books "Ocean Warfare" in an Indian Navy's submarine by Dr. Kalam, during his Presidential Fleet Review, embarked on a submarine sortie. It was a tremendous experience which made me understand how difficult the life inside a submarine is. Accidents may lead the submarine to sink and may also prevent the crew from escaping that sinking submarine.

Fitting BRAHMOS will give submarines extra teeth for combat and make them lethal. Currently, Indian navy is operating the German HDW class and the Russian Kilo class submarines, with French Scorpene submarines under construction. New class of submarines called P75 (I) will become potential power with BRAHMOS.

Installing BRAHMOS missile in these submarines will give an edge in the combat. Some of the ASCM are launched through the horizontal torpedo tubes of the submarine. Since BRAHMOS has a greater diameter than the torpedoes, launching it through those horizontal tubes is not possible. Nevertheless, the missile can be launched in that fashion. Also the usage of missiles in the place of torpedoes reduces the fighting potential of the vessel since the torpedo can be effective against both enemy submarines and surface vessels. Using missiles in this fashion will reduce fire power of submarine. Sending some surface vessels is a good option to hit the enemy's land based targets. But there is a considerable threat for those vessels from the air. Today our potential adversaries have competent air force and as I have said earlier, nothing can stand before an aircraft. So sending a submarine would be the most viable option.

Since the BRAHMOS is a universal missile there won't be any alteration in the missile. The nose cap will protect the missile from water entering the canister. The presence of the canister will save the missile from the seawater.

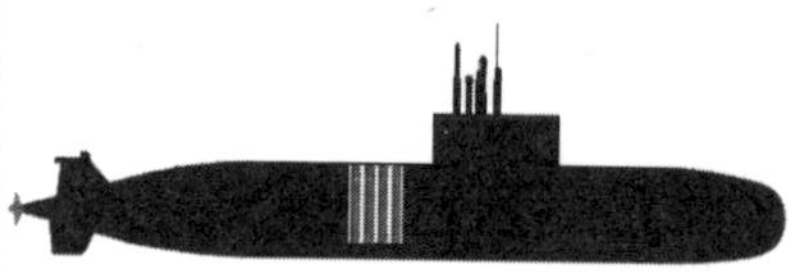

BRAHMOS in a submarine

The challenges in integrating the missile with the submarine are somewhat tough. The reason is that submarine is a vessel that sails underwater. As you may be aware, the pressure increases terribly underwater. Anything that has to survive underwater must be a hard nut. Now the challenge is to degut the hull of an existing submarine to integrate BRAHMOS section in the submarine. A slight miss calculation or even a small disproportionate work may lead to a catastrophe leading to the loss of the submarine and the crew.

This requires very complicated modifications to be made on the hull of the vessel with clinical precision. Also, enormous amount of simulation studies have to be carried out. Generally such work could be done only with the OEM whereas from our side we are ready with our missile and FCS.

Submarine launched BRAHMOS missile can get the target coordinates through a satellite or transmitted from a submarine

command centre. When the submarine gets the target coordinates, it launches BRAHMOS and can move away from that area. Similarly, the BRAHMOS missiles can also be launched to hit land targets. The underwater variant has been flight tested and is ready for induction.

SILO BASED WEAPON COMPLEX

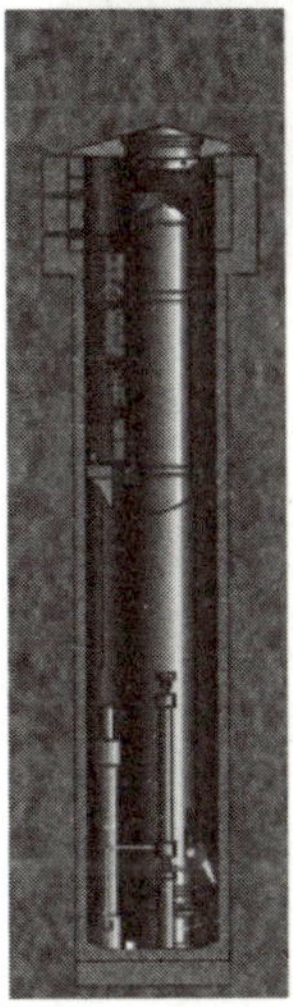

Silos are underground launchers that are constructed with hardened concrete tubes closed with a hatch. Missiles are kept inside these silos. Since these facilities are underground, they are difficult to destroy and can even withstand a nuclear attack. So, silos are strategic assets from which a nation can return a missile strike.

When requirement arises, the hatches are opened and the missiles are launched. During the cold war, many of the nuclear tipped ICBM systems of the both sides were inside these silos only. They were providing deterrence by assuring a mutual destruction second only to submarines with nuclear missiles.

BRAHMOS is also capable of being stored and launched from silos. A fuelled and combat ready missile can be deployed in a silo. Upon requirement, the missile can be fired on a short notice. When the missile is placed inside the silo, the scheduled check-ups can be performed without taking a missile out of the launcher.

These silos can be constructed and a weapon complex can be installed near strong points along the coastline. Such complexes can ensure destruction for enemy's seaborne aggression.

The silo based weapon complex will be consisting of:

- Silo based launchers.
- A Control centre from where the launch will be controlled.
- An underground or over the ground command post reinforced with heavy concrete layer.
- This command post will be connected to the higher echelons through various communication links.

A schematic sketch of a coastal defence scenario is given below. A similar weapon complex can be established for attacking land targets also. The enemy will also know that our missiles can finish the war

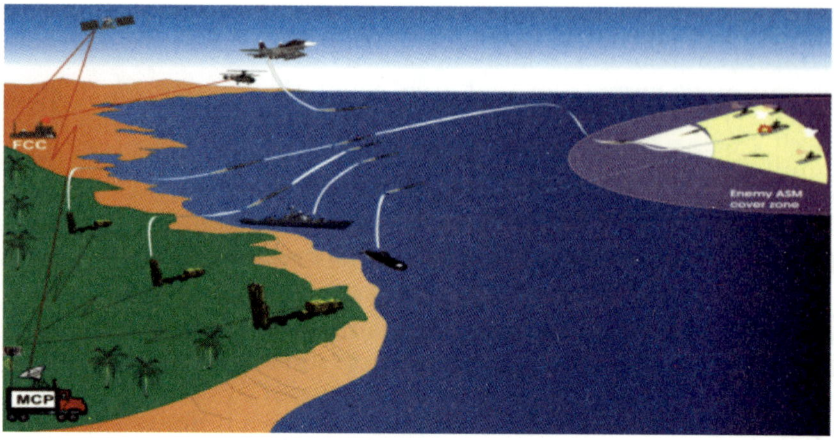

A coastal defence scenario

even before it begins. In that case, a non nuclear deterrent can be established with these systems.

INTO THE FUTURE: NCW CAPABILITIES

Warfare has been evolving into various stages and at each stage it is centred on some attribute. The very primitive form of warfare was human centric. Even up to some four hundred years before, the warfare was human centric. It was the number of the soldiers and their valour that determined the outcome of a battle. The army that had a huge manpower was considered strong. During Napoleon's era, the warfare turned towards 'mobility and marching.' An army that was capable of manoeuvre was considered strong; a manoeuvrable army can quickly encircle or separate the enemy. In the 1900s the warfare changed to platform centric warfare where superior weapon platforms like aircraft, warships and tanks, their fire powers and numbers determined the outcome of a battle. An army that was capable of using heavy firepower was considered the strongest. But we are in the age of transition to Network Centric Warfare (NCW). The future wars are going to be network centric wars. BRAHMOS exactly fits in the network centric architecture. But before that what is network centric warfare?

What is Network Centric Warfare?

An army that is highly communicated and informed is the one that is considered as a superior army. According to Sun Tsui,

"If you know the enemy and know yourself, you need not fear the result of a hundred battles; if you know yourself but not the enemy, for every victory gained

you will also suffer a defeat; if you know neither the enemy nor yourself, you will succumb in every battle".

NCW is somewhat similar. If we know that where the enemy is, what he is doing and also know where our force is, then winning the battle is imminent. How can we know about the position of the enemy and after knowing that what we can do? C4ISR is the buzzword, which is expanded to Command, Control, Communications, Computer, Intelligence, Surveillance and Reconnaissance. When these entities are closely interlinked, they result in the network centric capability.

In a typical war, first the offensive capabilities of the enemy and his defensive line must be assessed. But how these details can be gathered? For that the enemy has to be continuously monitored by employing surveillance assets. A typical example is an imaging satellite. Apart from that, a closer level reconnaissance must be carried out by employing reconnaissance assets like aircraft with imagers. By analyzing these data, intelligence is gathered and that is the knowledge on the enemy. After this assessment proper planning is required to attack the enemy at his weak points and to strengthen our defences at our weak points. The plan has to be executed in close coordination at predetermined times; so the execution has to be commanded and controlled properly from the command centres. The decision making command centre will be far from the frontlines. So flow of command to the frontlines from the command centre and flow of reports back from the frontlines to the command centre is necessary. A delayed report or command can change the outcome of the entire battle. So a proper communication is required to streamline the conduct of action on the enemy. After attacking the enemy, there should be 'Battle Damage Assessment (BDA).' BDA is also reconnaissance but carried out after an attack. This is to confirm the destruction of the target and if the target is not destroyed another wave of attack is carried out. Though, this is the normal procedure for the conduct of war, when all the above mentioned aspects are integrated in real-time, then such a capability is called as NCW. Integration can be done by employing various sensors like early warning radars, satellites, etc. Sensors are deployed also through military units like aircraft or naval vessels. In a nutshell, every single unit would be connected to the high command through the immediate command. The high command can access each and every individual unit in real-time. Such a robust communication link would enable improved information sharing. This information sharing would give a complete picture of the battlefield, which in military parlance called as Situational Awareness. This situational awareness will lead to better command and control and hence mission effectiveness.

Say for example, an early warning aircraft detects a huge enemy aircraft formation that has to be intercepted. Some of the friendly aircraft are also on the air returning from another mission. These aircraft cannot be directed against the incoming enemy formation. Why? Because the status of the friendly aircraft may vary, some may be out of weapons or fuel and some may be damaged. So the command must know the status of the forces and where our forces are. When the command knows all the minutiae about the friendly aircraft, it can choose the aircraft that are fit to fight, and can be sent to intercept the enemy. But how does the command know all these information? Here comes the role of communication links.

Importance of Communication Networks

Every layer in the theatre is networked with each other; submerged submarines, surface vessels, infantry & tank columns, aircraft and satellites etc. Though this looks simple, it is very difficult to realize since each and every military unit has to be connected. If there are 1000 combat aircraft in the air and 1000 infantry platoons operating on the ground, just imagine the extent of communication involved. The most complicated issue is that every class of military units operate in various frequency bands in the communication spectra. And even in a single band, the requirement will be huge chunks of voice and data communications and that too in real-time. So a side's superiority in network capabilities will decide the course of war. Like Sun-Tsui said in the NCW, 'the weapon is going to be information and flow of information.' The targets will be enemy's network and every network will have their own choke points, their vulnerabilities. So hitting at these points will damage or degrade the enemy's network to reduce his combat capabilities.

This capability will be achieved by many nations in another decade. Not lagging behind, Indian forces are also gearing up towards NCW. In that accord, BRAHMOS missile system is a pioneer in NCW element since the land attack versions for the army are purely operated in NCW principles.

BRAHMOS in NCW

If you look at IGMDP, the principle is to provide the user with a state-of-the-art weapon system at the time of the induction. So when we started the development of BRAHMOS we again wanted the system to be futuristic. During the time of development, the ideas of NCW were emerging in the horizon. So the BRAHMOS missile system was developed to be a node in the network centric capability of Indian forces. The land based weapon complexes; both the land attack and the coastal defence variants have been developed as NCW capable and they operate in that fashion. The present sea based weapon complex will automatically become an NCW element when the carrier vessel turns as an element of NCW. Similarly the submarine based and the aircraft based complexes will also turn as NCW element when the corresponding platform changes so. Let us see the structure of the weapon complex in NCW from the bottom to top.

For example, if a target is situated in a coastal city, the decision will be made on how the attack has to be performed. It may be carried out by a ship based weapon complex or from the land based weapon complex or from an aircraft based weapon complex. When decided to attack from the land, the instructions are passed on to the individual MALs operating from a point from where an easy strike can be achieved through the MCPs. Then the MAL will decide the route plan and then the missile will be fired. The reports will be sent from the MALs to the higher echelons through various encrypted communication channels.

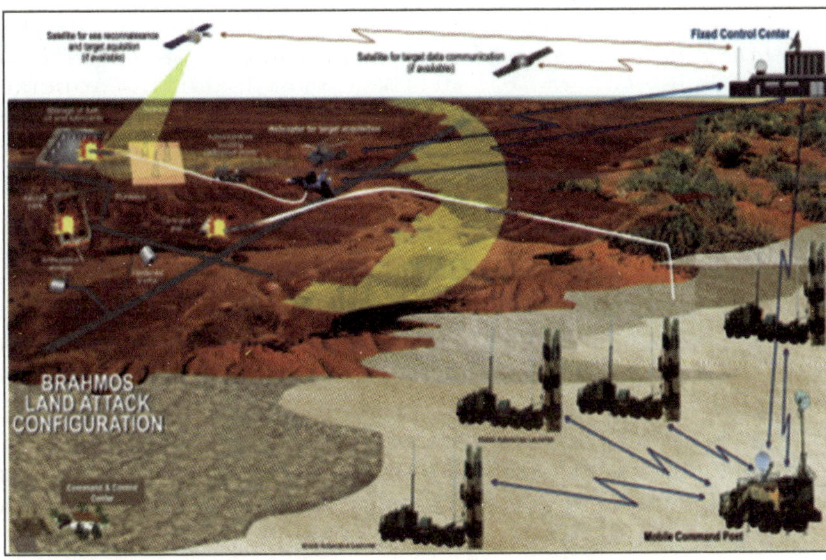

Schematic sketch of NCW with BRAHMOS in land attack role.

When the entire force structure is shifted towards the network centric capability, then the BRAHMOS systems will easily fit in the new networks that will be built. The figure shows the schematic sketches of BRAHMOS in land attack missions (NCW).

12

ENHANCING THE CAPABILITIES

Initially, BRAHMOS was supposed to be just an anti-ship missile, that too to be launched from the shores, through mobile launchers to hit enemy vessels deep in the sea. But during later stages various versions and variants started to emerge. Just from a single missile BRAHMOS graduated into cluster of missiles. This chapter is about that evolution of BRAHMOS as a universal missile to be launched from any platform, against land and sea targets, in any desired trajectory.

SHIP LAUNCHED VERSION

Any tactical missile that could be launched from land can also be launched from a warship but with some modifications. The challenges in this process are less, but decisive due to the harsh and corrosive sea environment. So we were confident about a ship launched version. After the completion of first flight trial, a series of studies were initiated. Our studies revealed that the missile could also be launched from an inclined position as well as from vertical position also. But the roadblock was that a vertical launcher suitable for ship was not available with us then. Though the missile is launched vertically, the launcher and the associated systems are not the same. So they were to be developed - a time consuming process indeed. The proposal was given to Indian Navy for a ship launched version. This version was not the vertical launch scheme but the inclined launch scheme. But to prove the missile's capability of launching from an inclined position, a test flight from a static inclined launcher had to be conducted. After a year from the first flight trial, we conducted the second flight trial that was from a stationary inclined launcher. But during the preparation for this flight trial, we were informed that that some spy satellites might be prowling over the Integrated Test Range.

It forced us to play the game of hide and seek; in the night time we positioned the missile at the launcher and the launcher was covered with tarpaulin. From an aerial view it would resemble like a tent or a truck.

This test proved that the missile could be launched from an inclined configuration. The success of the flight trial gave the confidence for the Indian Navy to go ahead with the installation of BRAHMOS in a frontline warship. This proactive decision of installing BRAHMOS in INS Rajput by Indian Navy, especially Vice Adm. John DeSilva, the then Flag Officer Commanding in Chief, Eastern Command gave a big boost to the induction of BRAHMOS in Indian Armed Forces. INS Rajput is a formidable vessel and was equipped with P-20M anti-ship missiles. We replaced the existing P20 Launchers with two BRAHMOS missile launchers on the deck in an inclined configuration. Within a year, installations were completed with launcher, Fire control system and reference Inertial Navigation Unit. The integration process involved physical, mechanical and electrical integration. The vessel was almost 2 decades old having many of the legacy equipments and electronic systems. Integrating a state-of-art missile with such a vessel was a formidable challenge. But our scientists and engineers met the challenge very well and the missile system was integrated with INS Rajput.

VERTICALLY LAUNCHED VARIANT

After successful installation of the inclined launcher with advanced Fire Control System built in India, several flight trials were conducted against the decommissioned ships as targets. These flights proved the reliable performance of the missile and the accuracy of hitting the target at every time. Navy desired to carry out an exercise of missile launch with live warhead against a decommissioned ship as target. The launch proved the devastating power of BRAHMOS to the Navy. Immediately decisions were taken by the Indian Nay to install BRAHMOS in number of "to be" constructed ships and those undergoing refit.

INS Ranvir was offered for fitment of BRAHMOS. This time a debate came between inclined and vertical launchers. Already ground tests have established the vertical launch capability of the missile. Hence, there was no difficulty to introduce in ships. But it needed unique design of Universal Vertical Launcher Module (UVLM) to have eight missiles capable to be launched in salvos. The proposal was welcomed by Navy as the vertical launch capability will allow the vessel to have more missiles in a combat ready state and also to give 360 degree coverage for the attack of the targets. We designed the new launcher system UVLM that would be same for any kind of ship. The launcher was successfully installed and the tests were carried out successfully.

THE LAND ATTACK VERSION

While we were installing BRAHMOS on ships, a thought process came to our minds that whether we can extend the capability to land targets for the Army. Even for ships they required hit coastal targets. We had

also proved at that time the launch from mobile complex at the vertical mode. A series of discussions took place whether such a novel concept of using an anti ship missile for a land attack is possible. Specialists, who had designed and developed the seeker, came out with a brilliant solution of software modifications and a new cyclogram to enable the seeker to home on to land targets. We decided to put the seeker on an aircraft and used the aircraft as a flying test bed to characterize the seeker and the software. We tried out this test bed against different land targets for the Army. We obtained extraordinary results which gave confidence to use the same seeker for both land and sea targets. It was a remarkable turning point for BRAHMOS. We conducted two flight trials which were witnessed by Gen JJ Singh, the then Army Chief.

Generally when any of the armed forces want to induct new systems they used to raise a capability requirement for the system. That requirement will say that what the system must do, in what battlefield conditions and in which type of terrain it would be used. Then they will ask the developer to develop a system on that basis or invite a seller to submit quotations in accordance with that requirement. This requirement is called by the Indian Army as "General Staff Qualitative Requirement (GSQR)". Being happy that BRAHMOS could be adapted for land attack roles, the GSQR was raised by Army. Development of MAL, MCP, and associated systems were kick started.

A dedicated team from Army worked with us in the development process. Their presence helped us to understand the field requirements of the user enabling the best use of the weapon and making it user-friendly. We introduced a jam proof navigation satellite receiver compatible to GLONASS and NAVSTAR. The innovative young scientists from DRDO, NPOM and BrahMos evolved an algorithm which can help the receiver to receive data from different global positioning systems and improve the accuracy much better. The new addition for the land attack system is the Mobile Command Post and the Fire Control System fitted in the mobile launcher. We had to incorporate various communication systems according to the requirement of Army. The successful implementation and evaluation of all the systems met the Army's requirement and the contract for the first Regiment was ordered. We delivered the deliverables well before the deadline and the Regiment was operational. This made the Indian Army the only land force to have the Supersonic cruise missile in the armoury.

Land Attack Version (BLOCK-II)

After inducting the land attack version, Indian Army wanted the missile to have target discrimination capability, i.e., the missile has to find and hit a specific target among a group of targets. Simply putting, the requirement was to hit a specific building among a group of buildings.

This new requirement of discrimination of targets posed a challenge to us looking for an innovative solution. So we pooled together the best brains from India and Russia to debate on the possible solutions including algorithms and succeeded in evaluating them in the laboratory tests. An in-house modification to the seeker was carried out. Similarly, a new navigation mode was developed. However, the first trial of the Block II version did not hit the target as the GPS receiver did not get signals from the satellite clusters. I will narrate an interesting story associated with this fiasco in the flight trial experiences. During the subsequent two trials, the target was accurately hit by the missile.

Land Attack Version (BLOCK-III)

While Block I and II versions of the missile are capable of neutralizing the targets on plains, the requirement for the Army extends to operations in mountainous regions. Sometimes, the target may be situated behind the hills. Such targets can only be engaged through a near vertical dive. So, the need came to evolve a new trajectory for steep dive mode attack. Initially, we had hesitation as the ramjet engine has to operate in the reverse direction maintaining the flow of the fuel. We carried out several ground tests and got convinced of a steep dive trajectory being possible. Steep dive gives full speed through the engine instead of gravity fall. Flight tests were conducted to demonstrate the capability.

SHIP LAUNCHED LAND ATTACK VERSION

As we were in the course of development process of the land attack version for Army, we planned to demonstrate to the Navy a ship launched land attack missile. Since we had initially conducted a flight trial from INS Rajput, we were having a platform ready with us. We located a suitable island target in Andaman Nicobar. Detailed study on two islands in Nicobar region was carried out for installation of the target. We had tough time to work in a remote island to construct the target and carry out the mission. Indian

Navy provided complete support including helicopters, ships and logistics. The launch took place and the target was hit very precisely. With this capability, Navy can strike enemy's land targets from a stand-off distance.

Continuous Capability Enhancement

PART 5

Sailing through Rough Seas

> **"There are no secrets to success. It is the result of hard work, perfection, learning from failure, loyalty and persistence."**
>
> **- Colin Powell**

The voyage was not smooth always. Rather we came across rough seas, which were difficult to endure. In this part, I am going to narrate some of our experiences wherein we had to put in all our efforts with ultra fine focus to overcome the difficulties.

13

EXPERIENCES OF FLIGHT TRIALS

In such a big country like India, only a handful get chance to work on hardcore missile projects and can get, feel and live with such experiences. Others would never get a chance to understand the hardships and nerve crunching moments that we faced during the conduct of flight trials. The experiences I am going to describe are unique and interesting for the readers. It is not that always we will be experiencing success. This is a game of risk. We must have the courage to face failures and turn them into success. Every step must help to move forward. Mahatma Gandhi said, "Strength does not come from physical capability. It comes from indomitable will."

Flight trial is the dynamic evaluation of various subsystems to perform in an integrated manner to achieve the set mission objectives. It is in a way, the culmination of development efforts on ground through intensive test and evaluation and proof of the hard work.

The missile has major subsystems like propulsion, structures, control, guidance, electronics, software and telemetry, along with power supply, connectors, junction boxes, relays, logic circuits, processors, many kilometres of cables and so on. Each unit has to go through qualification tests to the required environmental conditions specified in military standards. This rigorous testing and assessment of reliability ensures efficient functioning of the unit. Once such tests are complete and certified by the Inspection Agency, these units are assembled into subsystems which again get qualified for integrated performance on ground under different levels of testing. Then comes the interfaces between the subsystems which are ensured in the checkout complex. During checkout, electrical performances of individual subsystems and interfaces are cleared with specific conditions of approval outlined in the procedure. The units which are earmarked for the flight tests through similar procedure, but only to the acceptance levels of environment as specified in standards.

In order to verify the integrated performance as a system to meet the specific missions, the missile is "flown" on ground through Hardware-in-Loop-Simulation (HILS). Here, the flight of the missile is simulated by giving an external input to the sensor. Say for example, a sensor called as pressure gauge to monitor the pressure build-up in the missile. Pressure gauge generates more electricity when the pressure increases. The guidance system checks the increase in electricity and understands that the pressure is increasing. Let us take the example of electronic weighing machine to understand this. When we stand on that machine, the pressure gauge senses the pressure introduced due to our weight and generates electricity. When another person stands, it generates electricity according to the weight of the person. With that idea in mind, let us see the HILS. During HILS, the pressure gauge is replaced and a signal is injected manually, quite similar to connecting a small battery for electrical signal. Now, to simulate the increase in pressure, one more battery is connected. Similarly, there would be many sensors. In HILS, working of the missile subsystems is tested. This test validates that a particular subsystem is working according to the design. All this may not be sufficient. Several failure modes are also simulated to study the effect. The mission algorithm also gets proved in HILS. Then we are ready for the flight test of the missile.

What is required is to see whether the said subsystem works in the actual flight. So for each flight trial, a set of test parameters are fixed and then tested. Now, how do we know that subsystem is working while it is flying with the missile? So we need an electronic system which can transmit the missile performance to the ground. This is called telemetry system. So the role of development flight trials is essentially to check the performance of the system. Onboard Telemetry is needed to record the performance of the subsystems and ground units to receive the data, large computers to process the data and to give real time analysis of whatever is happening to the missile. During the flight trial, we can see a real time trajectory plotted in front of us in the Control Centre, thus giving the actual performance of the missile.

All these developmental flight trials are observed by a group of personnel from the armed forces. Once they are satisfied they give a set of simulated combat scenario. This is called as user defined flight trials. These flight trials take place according to the combat requirement of the user. Once the objectives are met, i.e., the target is hit under simulated combat scenario, the system becomes ready for deployment. In addition, an elaborate training of the Army Personnel is carried out to ensure the readiness of the system for actual deployment. For further assurance, the field trials are carried out by the trained crew from ship for Navy and from desert or in the mountain region for Army to conclusively decide that the system is ready for combat.

After induction into the armed forces, the users conduct the flight trials themselves. Those flight trials are termed as user flight trials. During these flight trials, the user would be the one who would operate the systems and the scientists, who developed the system, would be supervising it. The results of such trials give confidence to the user for war preparedness. The deployment process starts.

From the above, we realize how important a flight trial is. Let me describe the action scenario during the countdown of a flight trial.

WHAT IS COUNTDOWN?

Most of the readers would be aware that a counting sequence is done in the reverse order during flight trials through movies and video news clippings. But 'why it is done?' and 'what is done during that time?' is not generally available in the open domain. I will describe to the readers a typical countdown process for Aerospace missions whether it is ISRO or DRDO or NASA. For ease in representation, I will use the term missile.

Missile is highly complex system with thousands of subsystems which may work individually or sequentially or as groups. A sub system may be required to work only for one second. Say for example there could be a valve whose functionality may be is to open and close. Simple in words isn't it? But the real complexity is its timing. That valve has to open at the exact second. Its opening a second before or later may even lead to catastrophe.

Then comes the human factor. It is a mission and therefore several agencies are involved. The place from where the missile launch is to be performed is called Missile Test Range. This range needs to be ready with all safety clearances to give go ahead for the launch. A flight NOTAM (Notice to Air and Marine Movements) is issued to avoid traffic during the launch. All the tracking radars, telemetry stations, computer network, safety operations are networked and put in operations in the range. The naval ships are deployed near the target points with telemetry stations on the ships as well as helicopters to acquire the data on the target hit. Also, at the launch point, a number of telemetry and tracking systems including radar and electro-optical systems, data processing centres, mission control centre are all activated in a network form. The integrated coordination and the proclamation of the count-down sequence are carried out at mission control centre. For each mission, its overall control is governed by the mission director who integrates the missile, ground complex, range systems, meteorological data, range safety, down range stations, communication network, processing centre and emergency operations.

Once a flight trial is planned, the first parameter to be checked is the viability of the meteorological conditions required for the missile launch. It is said that during the Cold War, both sides would wait for the adversary's spy satellite to pass over the test range. Ah! Not forgetting to mention, it is not only meteorology but some even refer to astrological conditions as well, especially the star features of the Mission Director.

Next, the time is fixed for launching the missile. Then we calculate back and set the clock for countdown. This time period is called the preparation time for the missile to be ready for launch. Each and every subsystem involved in the mission whether in the missile or on the ground is thoroughly checked and made ready for the launch. Say for example, our PSLV requires a preparation period of 60 hours. There are systems, which require even 96 hours of preparation time.

If the missile is liquid fuelled rocket engine, like Prithvi, then preparation time is required for fuelling. It is to be noted that the fuel being corrosive in nature, is not filled beforehand. Once it is fuelled, then the missile has to be launched and cannot be stored beyond a certain time. Fuelling is a process which has to be carefully carried out to avoid any catastrophic explosion, as the fuel and oxidiser used are hypergolic (instantaneous explosive on contact). In the case of PSLV, GSLV and also Agni (for its velocity trimming module and control), the countdown time includes fuelling of propellants. But solid propellant based missiles like Nag, Akash do not have such requirement and can be quickly deployed. Since BRAHMOS has storable aviation grade fuel for its ramjet engine and solid propellant booster, there is no such requirement to be included in the count down.

There are many such activities that fall under the preparation time. All these activities are managed through coordination of time and space. Here coordination may precisely be termed as synchronization. The reverse countdown synchronises all these actions. While describing the count-down, I thought of an interesting episode to share with the readers.

My son Dr. Bhagavath Kumar is a Surgeon, who after his specialization M.Ch. in Plastic Surgery at Stanley Medical College, Chennai went to

USA for fellowship in Micro and Hand Surgery. My grandson, Sivathanu Kumar who was only five years old at that time, telephoned me from USA and asked "why in India, they taught me to count from 1,2,3, ... 10, whereas in USA, they start with 0,1,2,3, ... 10." I said since, '0' has no value, we start counting from 1. He caught me by saying, "Grandpa, you showed me the video of launch of a missile. When they get ready for launch, they say 10, 9, 8, ... 1, 0. When they say 'Zero', then only the missile leaves. So '0' is important." Indian schools should also realize the value of '0'. He bowled me out straight. I felt that Indians who found out Zero, should count up from 0, 1...

Generally, the exact instant of the launch is called as T_0(Read T Zero) regardless of the time in the clock. If there is 30 minutes to launch then that time is called as T_0 minus 30 (T_0-30) minutes. After the missile is launched, the time is counted in normal sequence. So, 2 minutes after the launch is called T_0+2 minutes. During the countdown all the actions are executed in a coordinated sequential manner. Each group confirms that they have executed their share of task. Tracking stations reach the area of responsibility some hours before the launch. So everybody will confirm that they have executed the action to which they are responsible for.

Before the count-down, all critical parameters of the launch, team assembles and conduct *Pooja* at launch point and break coconuts. They take the respective position. Mission Director announces the start of the count-down. As the count-down progresses, every system and operation gets confirmed for readiness. Mission Director after satisfying himself authorizes launch on auto mode. During the last few minutes, typically To-6 minutes, the automatic check out system takes full control. The check out system gives the signals for various operations for full readiness of the missile and checks the parameters for correctness. The last ten seconds are like *'agni pariksha'* (world's toughest test) and the heart rate increases to abysmal rates. The authorised fire control officer presses the fire button at the destined time. The missile gets activated and takes off. Then, we start counting up and announce the parameters being displayed on the monitor till it hits the target. The flight trial team can instantaneously tell that the flight trial was successful. The Mission Director announces the fulfilment of the mission objectives and thanks all the participants. Within the next two hours, the post flight analysis team presents quick look results to confirm the performance.

I have participated in flight trials from land and sea from the days of SLV-3 to Agni to BRAHMOS. It is full of tension as the launch campaign team never want their hard work to fail. The flight test can be a success or failure, as there are more than hundred thousand components to work as expected. Discrepancy in performance of a single component may result in mission failure, even though every subsystem is checked thoroughly and prepared with enough care. Once the mission succeeds, it is a great sight to see the faces of all the personnel who were involved in the flight trial operations; same is the situation with the observers. The faces shine bright like sun while tears of happiness ooze out like drizzling in the sunshine. Flight trials trigger emotions on unimaginable scales. Now let us see some of our notable experiences.

FLIGHT TRIAL LOCATIONS

Generally, missile flight trials are conducted at Integrated Test Range (ITR) Chandipur at Sea near Balasore, Odisha. It is a beautiful place for tourists. The Bay of Bengal recedes four km during low tides and the waves will be in full blast during high tides. This happens twice in a day and is an interesting sight to watch. During low tide one can walk on the sea with bare foot and collect shells (Sea weeds). The British have located this site for trying out firing of shells (Ammunitions) from artillery guns. This establishment for evaluating the Armament Shells is today called Proof and Experimental Establishment (PXE) and has now completed more than one hundred years.

During IGMDP days in 1980s, we needed a range for testing of missiles. SHAR facility of ISRO at Sriharikota was used initially for the Devil, Trishul and Prithvi Flights. Obviously, the DRDO needed a missile test range of its own. Balipal in Odisha coast was identified as the National Range. Extended PXE became the Interim Test Range (ITR). Later, the availability of Balipal range became an issue and therefore, the Interim Test Range became permanent and re-named as Integrated Test Range. Whenever we launched Agni, we needed to evacuate 3500 villagers from the nearby villages in order to have the required safety clearance. Though villagers oblige without any problem, it was posing considerable difficulties. We needed a better safe range. That was the time we were also looking for a land target for Prithvi flights, as we had to prove the accuracy of impact to the Army for their acceptance. The Wheeler Islands off Damra could give a 70 km range from ITR. Once the launch took place successfully and Prithvi landing very close to the target put on land, Dr. Kalam decided to use the Wheeler Island as the launch pad for Agni and its future bigger missions. He met Mr. Biju Patnaik, the then Chief Minister of Odisha and asked for the site. The highly patriotic Biju Patnaik immediately agreed and suggested that, "Dr. Kalam you must launch ICBM from Wheeler Islands and make our country Strong". Today,

Agni-V launched from the Wheeler Islands fulfilled the dream of Biju Patnaik and Dr. Kalam.

The test range has a wide array of instrumentations like tracking radars in multiple bands, electro-optical tracking systems (a kind of camera) in visible and Infra Red lights, Fixed and mobile telemetry stations, tele command stations, range safety, meteorological stations, real time data acquisition and process computers, camera stations, etc. This facility became state-of-the-art facility for testing missiles and aerial vehicles. We utilized this location very well and all our developmental flight trials were conducted from here. Similar facilities are available in the Wheeler Islands and its nearby shore.

The other location is situated in the Pokhran, Rajasthan. It is called as Pokhran Field Firing Range (PFFR). This area is extensively used by Army and Air Force to practise firing of their equipment and weapons for their armament evaluation. We used this range for proving our warheads used for Prithvi. The other range is ship. Navy carries out their trials at Sea, with ship as the launch pad.

TESTING THE RUGGEDNESS – RISK TAKING

When we were preparing for the first launch at ITR we started loading the missile in the static vertical launcher exclusively erected for the BRAHMOS system. For loading the missile inside the vertical launcher, it was necessary to lift the missile to double its length and slide it inside. We were doing this process at the launch pad. The missile was being lifted through a crane with a special lifting device holding the nose cap. When the missile got lifted 15 metres above the ground level, the holding device inadvertently yielded and broke. The live missile fell down, back on the corsette – the transport trolley making loud noise. It was a pathetic scene and the operating personnel ran away for safety. A live missile is terribly dangerous even if it is without any warhead. So everybody's breathing stopped and slowly some of them gathered courage to approach the missile for observing any abnormalities. We got the information and swiftly acted to isolate the live missile for detailed examination by experts. Russians friends were very upset by the incident and advised me to destroy the missile by following a procedure. This is what they do in their country.

For the flight test, we had prepared two missiles, one is called main and the second one is called stand by. Having got into problem with the main missile, without loosing time, the stand by missile was loaded after corrective measures were taken for the loading fixture. We got ready for the launch and the mission was successful. I will narrate the details of the first mission later.

After the successful first launch, we studied the fallen missile and made an elaborate plan to check each sub-system very carefully and analyze the results. We were satisfied with the health of the missile and thought of going ahead for a flight trial instead of destroying it. Of course, the specialists also went through our study and later declared that there was nothing wrong with the missile. Still a dropped missile cannot be flight tested as we do not know the possibility of any internal failure. But the Indian team took this as a challenge and after a careful review, we decided to carry out the launch. There were apprehensions from the Russian specialists, but we were convinced based on the results of the tests carried out for various subsystems. Finally, the Russian side also agreed to carry out the launch, when we presented the details of the tests and results for every subsystem. The Indian side took the risk fully. We formed the Flight Readiness Review Team and thoroughly looked at all aspects. The Launch Authorisation Board (LAB) cleared the launch. The launch (FT 02) was 100 per cent successful and our scientists were even more jubilant than the first successful flight of a healthy missile. Such was the robustness of BRAHMOS. But, what was more important was the decision making process. The risk was analyzed carefully and the boldness shown to face the risk paid off immensely and proved that the missile is rugged.

LAUNCH OPERATION FROM MAL - COURAGE

The relationship between men and machines dates back to the ages. No machine is complete unless the man using the machine is competent. But as the military machines are used in adverse and frequently changing dynamic situations; the man operating it should also be highly competent. The outcome of such integration not only affects the man and machine, rather it plays a vital role in upholding the pride and honour of a nation. The man behind the machine plays a significant role in exploiting the full capabilities of the war machine. But human interface with the war machines is the last in the chain of proving a system. This is because the safety of the system has to be proved by successive successful trials.

The missile was to be test fired from ITR to validate its operation from a mobile launch complex with an advanced guidance system and manoeuvring trajectory meant for the Army. Lot of activities were on full throttle in the preparation of launch from Mobile Autonomous

Launcher (MAL). Army had posted a specialist team meant to take over the missile group, spearheaded by Col. Bishnu Ram.

I still remember the indomitable spirit of our Colonel to launch the missile from the MAL itself himself. As I have said earlier putting a man behind a lethal machine is only kept at last but the courage displayed by the Colonel was astonishing, even for the first launch from MAL. The missile once launched is going to fly over his head, with roaring sound.

He argued, "Sir, my organization has placed me here with some purpose. This system belongs to us and I have grown with the system. I must launch the missile from the MAL as per the ultimate configuration of design." I was overwhelmed with the courage and faith of the officer on the system. He showed the finest traditions of Army "Displaying selfless devotion and leading from the front."

Normally, till the time the system design is frozen and development is complete, such an adventure is not advisable. Safety concerns were alarming as the missile was to fly over his head. Missile was loaded with hundreds of kilograms of propellant. Any mishap would turn into a catastrophe and such a thought still brings out chill in my spine, since failures are part and parcel of flight trials. Even developed nations like USA and USSR have faced failures in flight trials and have lost valuable lives of their astronauts and cosmonauts. You might be aware that Yuri Gagarin was the first man to reach space. He was a test pilot and he died when he was test flying an aircraft. So, accidents can happen at any time. We don't forget our courageous astronaut Kalpana Chawla, who flew in Space Shuttle died.

But confidence of the Colonel on the system and his courage to accomplish the task edged us to go ahead to launch the missile from the MAL. Although, I knew the worth of the system, but the thought of one in million possibilities of a failure pumped my adrenaline high. The head of FCS division VSN Moorthy also volunteered to be a member of the fire team. Here, I have to laud his guts and grit. Normally a scientist thinks twice to take risk. But he is a different scientist - calm, quiet and firm. He volunteered himself to be along with the Colonel who is a soldier to fire the missile from the MAL. We asked Russian friends who had more experience in handling similar systems. But they did not volunteer and advised us for a remote launch. We somehow convinced them that we will go ahead with the launch with our men inside the MAL. I spent sleepless nights pondering the issue of MAL launch with our men. I was worried because it was the first such experience for our missile team. If any mishap occurs, it would be a disaster and the end of BRAHMOS induction in Army. But the sheer grit and courage shown by our men was negating the risk and helped us to overcome the worries. We mounted a big contingency plan at the launch pad to face the worst.

The destined time came. After prayers, we wished both men good luck. They stepped inside the MAL at the scheduled time. I remembered the Apollo 11 mission in July 1969, when Neil Armstrong, Michael Collins and Edwin Aldrin walked inside the Lunar Module LM 5. In the same style, our friends walked inside the MAL. Both officers were calm showing no sign of hesitation, rather they were smiling. The launch pad was evacuated and we settled in the blockhouse in front of the consoles. CCTV cameras were inside the FCS cabin of the MAL to monitor the crew conditions. The crew with their smiling faces comforted us and the countdown was initiated. The atmosphere inside the block house was tense. Our lips went dry as the final few seconds came! Till that time there was no glitch and everybody sat with their fingers clenched and crossed including myself. An emergency vehicle was waiting outside the Block House for me to rush to the launch pad, in case of emergency. The teams were alert to face any eventuality. The countdown reached the final ten counts.

10...9.....8....7......6.....5....4.....3......2.....1...... Zero!

The command was given. An ear breaking thunder rose. Some of you might be living close to the airports and would complain about the noise of the flying aircraft. But the level of noise that a missile launch creates is exponentially higher and is hard to endure when compared to that of a flying aircraft. Amidst that thunder, the missile came out of the canister like a shining sword coming out of the scabbard. It rose high, rolled, pitched and started its supersonic journey right above the fire team. The blockhouse broke into applause and tears of joy. Yes, the launch was a success! The capability of the MAL was proved beyond anybody's doubts. I rushed to the launch pad to look at the courageous fire team, who emerged victorious without a single scratch. Their victory smile still shines in my eyes and will remain forever. I recalled to them the words of Neil Armstrong, "It is a small step but a giant leap for mankind." The fire team was so thrilled to be the first ever crew to fire a supersonic cruise missile for the land force. When inducted, I told them, Indian Army will be the first ever land force on the earth to have land attack supersonic cruise missile.

I will always remember Col. Bishnu Ram who commanded the first ever BRAHMOS regiment for the Army and Mr. VSN Moorthy a DRDO scientist who is now the Project Director of BrahMos, for their courage to operate the system. Our Russian counterparts were speechless in the admiration of the courage of our fire team. These men have certainly made their proud place in the history of BrahMos for firing a supersonic missile over their own head. Now, it is a routine practice that all our army version missiles are launched from MAL. In recent times, a young lady scientist Ms. ANS Prasanthi showed her guts by conducting the launch sitting in the MAL cabin along with the Army crew, thereby proving the user friendliness of BRAHMOS thus giving confidence to our armed force personnel. The same MAL will be used by Indian Air Force and next by Indian Navy for Coastal Defence.

A CYCLONIC TRIAL - DECISION MAKING

One of the flight trials was conducted at ITR in the month of June, the South West monsoon period. We intended to launch the missile to test some of the crucial parameters including advanced software and guidance; a sort of 'Agni Pariksha' for the missile. But unknown to us, Mother Nature wanted more from the missile. She wanted the missile to prove its all-weather capability. All the requirements were met and launch procedures were carried out systematically. The missile to be tested rolled out as first of product fully prepared by Indian Scientists at the production facility. It was transported to ITR through road and checked out for readiness at ITR. The Mobile Launcher with the Fire Control System, Mobile Command Post and all the Peripheral systems were made ready. But we couldn't notice the smile of the nature on us. Out of nowhere a severe tropical cyclone had formed in quick succession in the Bay of Bengal and was supposed to cross somewhere around the eastern coast. When we reached the ITR, the sea was rough and roaring to its full might, a scary sight of course. You might be aware that the wind conditions and the sea state during a cyclone are not at all favourable for launch of missile. The wind was in its full valour, whistling and roaring with a speed of around 150 km per hour, as reported by ITR's meteorological set up.

We follow a set of procedures to clear the flight tests based on the review by experts and those who have provided the sub-systems with the participation of user agencies, range and quality assurance groups. This type of review is called Flight Readiness Review (FRR). Based on the review and recommendations of FRR, the launch gets authorized. We were in the middle of FRR meeting on the Cyclone day and debated whether to conduct or abort due to the met warning. In the meeting, there were two schools of thought – one supporting the launch and the other opposing it. The reason was obvious as there was no need to take risk due to heavy winds and chances of heavy rain in the next morning when the launch was scheduled. The outcome of meeting was to wait for a minimum conducive weather on the next day and if it does not, to pack our bags to Hyderabad because the meteorologists reported a storm approaching. Though we went to bed, we could not sleep due to the haunting dreams about the next day's flight trial. The forthcoming day could prove to be an ultimate test for all the men and the machine. The missile had to prove itself with tall mission objective and which would be a proof for us. The Rain started to pour as if a chorus to the whistle of the gusting winds. I was not able to sleep due to the anxiety and opened the windows of my room from where I was able to see the sea. Indeed it was a fearsome sight to see the roaring sea. Like the waves of the sea, my thoughts were also oscillating and I discovered that my colleagues were also spending sleepless nights.

Why I am stressing over this is due to the fact that war is not a cricket match which can be called off due to rain or its result being decided with the Duckworth Lewis System. War may erupt at anytime and weather is an important feature that militaries take advantage of. Think of a similar situation on sea when the ships are subjected to heavy roll and pitch in different sea states. But the men in uniform on the ship have to fight the war. They have to launch the missiles to attack the enemy ships. So all the fighting forces, men or machine must be in fighting fitness even in adverse conditions. Our missile and systems are no exception to this rule. They are designed to face such adverse conditions. One cannot escape proving its system in the adverse conditions. But at the same time, one will not take risks during development tests to prove new objectives.

Next day, early morning, I went out to have a look at the sea waves and the fury of the wind. To my surprise, I saw all our people participating in the campaign were also at the sea shore enjoying the sight of the cyclonic sea. I went to the new guest house where the Russian Specialists were staying. The door could not be opened as the wind from the opposite side was blowing in full throttle through the door which was kept open by them. All our Russian colleagues were in the balcony looking at the sea. After I called them, a gentleman tried to open the door but the door was not opening due to wind. He again asked us to push the door from outside. The door still didn't open. So the people inside the room joined hands to open the door from inside. The culprit was the door facing the balcony. When that door was closed, the other door could be opened. When I went inside, I saw this strange group and one of them exclaimed, "Nature treats everybody equally!" Ripples of laughter started shaking the room due to this timely comment. This funny incident helped to reduce the flow of adrenaline to an extent.

The situation further became grim as gale winds were blowing. But to our hope there was a sudden clearance in the weather though it was not conducive. It was raining like anything; imagine how the cyclonic rain together with gale winds will feel in a coastal area. The current weather status was the only concession that 'Mother Nature' could offer for our launch. We did all the preparations in quick succession. Vice Adm. Arun Prakash, the then Vice Chief of Navy joined us to witness the launch. As the user arrived, we saw a sudden patch of light as the cloud disappeared, as his name, he brought 'Prakash'. The sun appeared. The Admiral brought smile to us. We acted fast and took the decision of going ahead with the launch.

The countdown reached its most crucial last phase. In that short period of time, we felt as if we are going to receive the question paper of some practical examinations; students of science and engineering disciplines can understand the problem very well! Though the test was for the missile, but actually the test was for the men, materials, labs and industries. Even for the experienced professionals, the last phase of count-down is a test for patience and endurance. Wind started increasing its speed and a warning came that a cyclone may be crossing through that area very soon. The mission was held back from launch as the Range Director did not consider the situation conducive for the launch.

The wind increased its speed, it was a point of decision whether to conduct the trial in high wind speed or to call off the mission. I said to my team to go ahead with the launch and not to worry about the wind speed, which has just reached the extreme limit of the design. The whole team including the then Project Director Venugopalan was prepared to take the risk. The hold was lifted. The count-down started.

We could feel the adrenalin rush as the beatings of our hearts kept increasing. Finally, the countdown reached zero and the missile took off. It rose up into the sky and then carried out its journey in a textbook performance. The take off picture clearly showed how severe was the wind. Even to move the exhaust of the booster to one direction was a herculean task. We realized that the missile was so robust in its control system design and structure that it could withstand such severe weather conditions. Especially, the inertial navigation system of DRDO technology performed beautifully. It was a great experience of decision making with risk. I got a feeling that "God is a partner to the BrahMos project". Similar feelings came to me many

times as the God has decided to see the success of BRAHMOS.

The famous British poet Alexander Pope once said,
Nature and nature's laws lay hid in night;
God said, "Let Newton Be" and all was light.
Like his words, the performance of the missile even in adverse weather tore open the screen of pessimism.

The astonishing performance of BRAHMOS was understood by Adm. Arun Prakash and other user friends who witnessed the launch. This boosted up the user confidence tremendously which resulted in a quick letter of intent for BRAHMOS in multiple ships for Navy, signed by Vice Adm. Arun Prakash, the then Vice Chief of Naval Staff.

ANDAMAN LAUNCH – OUT OF BOX THINKING

An Island Please!

We had to prove the land attack capabilities of our missile launched from a ship to a land mass having a target. This will enable Navy to use the missile for coastal targets. Therefore, we decided to conduct a flight trial in the Andaman Islands, as it was not possible for us to locate land target in the main land of India. Lt. Col. Anil Misra, another hero and I planned to carry out an aerial reconnaissance of the 572 islands of Andaman and Nicobar to locate our island for positioning a target. We got the help of Navy and the Combined Command of Andaman & Nicobar (A&N) who were excited with the proposal. But everyone did say that it would not be possible to get the environmental clearance from the Prime Minister's Committee to carry out a missile launch on a land target at Andaman & Nicobar Islands. The reason was presence of rare species of flora and fauna in most of the uninhabited islands and those occupied by the tribal natives. Still, we decided to first conduct the survey and choose the appropriate island from mission point of view to position the target for the BRAHMOS missile to be launched from a ship sailing in the Andaman Sea. It was a wonderful experience to see all the beautiful islands and the Andaman Sea from a naval aircraft. Though, it was not a joy ride but a mission, still we were unable to resist admiring the beauty of the Islands. We flew over the volcanic Baren Island which was first thought of as installation place for target. But it was not suitable for our mission. During the recce going South of Port Blair, one of the Islands of Nicobar got our attention which was Meroe Island. Adjacent to the Meroe Island, there were two more islands Trak and Treis. After the complete survey, Col Misra and I met the

Forest Range Officers, the Chief Secretary and the Lt. Governor to put our demand for Meroe, Trak and Treis islands to be allotted to BrahMos for its missions. Everyone got excited with the type of missile and the mission. They wanted to extend their support for the requirement. I could convince them that the missile is so accurate and the target will be positioned in the large sandy area leaving a small portion of rock and few coconut trees. I assured them that even a single coconut tree will not be affected due to the launch. We wanted to ensure that the target location is decided based on the ground reconnaissance on these islands.

Lt. Governor got excited with the suggestion and boldness to ask for the islands to be allotted. He asked me to come to his residence that evening for a tea, with a laptop containing presentations. I went to his residence that evening. The Lt. Governor received me at the living room and I saw many snacks items including my favourite items – Madras Mixture & Murukku. Lt. Governor said that first you show the BRAHMOS launch video again. I want my wife also to see. I showed the launch video. His wife exclaimed that what a wonderful missile. India is safe. She told the Governor – you must fulfil his aim to carry out the flight tests at A&N Islands. We got preliminary green signal from the Administration.

Adventure to the Island

As a step further, I detailed a team for ground recee. On the ground, the idea of undertaking foot reconnaissance was quite adventurous. The team comprised of Colonel Uniyal, Lt. Col Anil Misra, AD Rane and Joint Managing Director Alexander Maksichev. You might have heard that sportsmen, especially cricketers being highly superstitious. But the missile scientists are more superstitious than anybody else. Coming back to the story, the team looked for the auspicious time and sailed in the Navy's survey vessel INS Sagardhwani. Unfortunately, the sea was very rough and it started pouring heavily.

Everybody was courageous on the surface level but the sea conditions were sending shivers in their spines. They rang me up from the ship and informed the conditions to seek my decision to call off the journey to the island. I asked them to proceed further without fear as everything would be alright, because I had a firm belief that God was also our partner. When they approached closer to the island, two of the members got into a 25 HP engine fitted on a small Gemini boat which is normally used by the naval divers. Along with the team an executive officer and two expert divers from navy embarked their voyage. This might appear comparatively easy for those who were spectators, but in reality, it was very difficult when the sea is rough.

With the roaring engine, the Gemini started its journey from the ship towards the island. It was still pouring and misty. From the ship, the boat looked like a fading dot appearing and disappearing between blue waves. When they were 10 metres short of the beach, high waves toppled the raft throwing all the 5 members on the beach on their heads. Two members reached the shore with great difficulty but with scratches all over the body, as the shore was rocky. Immediately the navy team dashed back into the sea to get hold of the Gemini which was being snatched away by the waves. Again they were thrown away by a huge wave. Somehow the Gemini was pushed into the sea to fetch the second set of members. But it took 30 - 40 minutes because of the sea condition.

Now, on the ship, the second group of landing party was preparing to board the Gemini and they jumped into the unstable boat. Hundred metres before the beach the diver requested all onboard to help him in holding the Gemini to avoid a similar incident as in the first trip. This group was unaware of the hazards of beaching the Gemini. They jumped on to the beach by holding the Gemini from opposite ends But in a fraction of a second the Gemini was on top of them and the inner instinct of survival made it possible for them to immediately come out of the dangerous situation. After exertion, the Gemini was somehow pulled to the beach and obviously even they could not escape bruises and scratches. The fact was that one of them did not know swimming and he had to somehow beat his hands to move to the shore. It's all God's grace. After regaining their orientation and getting over the initial shock of head-banged landing, the team started their tasks.

Pushing behind whatever happened, the team started recce of the island. They made an effort to go over the entire island, which was however not possible either due to rocky face of high banks or due to dense forestation. Even walking on the beach with drenched clothes was tiring as the foot sank by 4-5 inches in the sand. Astonishingly, all the electronic systems which accidentally got soaked with sea water and sand, worked well! Unluckily, the drinking water which the crew was carrying was also lost so they were left without water. With bruises all over and the body drenched with salty sea water I am still unable to imagine how these men would have done the ground work over there with irritation all over their bodies. Irrespective of these difficulties, the team located ideal site over a large sandy portion for erecting the target. The location was decided taking into account that the missile hitting the target would not affect the ecology of the Andaman & Nicobar Islands.

Getting the Islands for BrahMos

After the detailed survey of the islands, we made our plans for a possible launch site from the sea and a place to position the target. But we needed permission for carrying out the trials, as the Government had to agree for giving those islands to be used by us. It was a great story. The administration changed by then. Lt. Gen. (Retd.) Bhopinder Singh was the Lt. Governor of the A&N who was earlier the Military Secretary to Dr. Kalam when he was President. We knew each other very well. So everything went on top speed. The Lt. Governor, Chief Secretary and the Forest Officers all were supported the mission unanimously. We got all the clearances in record time. We assured them that there won't be any damage to the flora and fauna of the islands because of BRAHMOS, as we were sure about the accuracy of BRAHMOS in hitting the target. Everyone was thrilled and wondered if such a mission was possible. Thanks to Col. Misra, our hero who toiled to get these islands for BrahMos. Then it was necessary to get clearances from MoD and Government for conduct of the launch.

Alas Tsunami Struck

On 26th December 2004, Tsunami struck Indian Ocean as a result of an earthquake with an estimated energy released on the earth's surface at 1.1×10^{17} Joules which is equivalent to over 1500 times that of the Hiroshima atomic bomb. The Tsunami hit the islands with a magnitude of 9.0 with an epicentre off the west coast of Sumatra. The violent movement of sections of the Earth's crusts called tectonic plates displaced an enormous amount of water, sending powerful shock waves in every direction. The massive Tsunami devastated the flora and fauna of the Nicobar region. The saline water had completely damaged the greenery of the entire region and some of the islands got submerged in water. Tsunami inflicted heavy damage to the Andaman & Nicobar Islands. The northern islands lifted up by

Lighthouse in Indira Point before and after Tsunami

1.5 metres and southern islands went down by the same height. The photographs of Light House in the very famous Indira Point in the Southern most point in Great Nicobar Island clearly depict the devastating effect of Tsunami.

It was a disaster which held back our mission. We came to know from the A&N administration that an Indonesian ship got stuck close to the Beach of Meroe Island due to the Tsunami. Col. Misra and I again undertook an aerial survey to inspect the fate of our islands and to decide over the launch. I saw the watermark of the islands in the Northern portion which lifted up and the light house at Indira Point down in the sea, and also the heavy damage to all the trees of the islands with sea water logging inside. We went to Car Nicobar and met our brave Airmen. They forgot all the agonies and started talking about BRAHMOS induction in Air Force. Lot of damages had been inflicted to the central portion of Meroe Island and sea erosion in Trak and Treis. We decided to shift our target point to the sea shore of the Trak Island. We took out Meroe from our consideration after seeing its condition and to avoid the foreign ship that got stuck. A&N Administration told us that the abandoned ship could become a target for BRAHMOS after one year if it is not recovered by the owner. All preparations started for the launch from INS Rajput in the Andaman Sea towards the target which had to be erected at the Trak Island.

An Enemy in the Distant Land?

A five member team consisting of Col Misra (BrahMos), Rane (DRDL), Dhir (TBRL), NS Srinivas (RCI) and Satyanarayana (RCI) had to construct a target on the island, make necessary alignments of the target taking into account the trajectory of the incoming missile and to stiffly place high speed cameras at suitable locations to capture the target destruction. A stand had to be constructed with metal rods. So it was decided to land the team on the island through a helicopter. For dropping and supporting our team, the Indian Air Force came for our help and they deployed one of their helicopters based in Car Nicobar. The helicopter was hovering over the island. When the team was about to get down from the helicopter, they saw some mysterious tracks similar to that of a truck. The team asked the helicopter crew

The Mysterious Tracks

about the tracks. The pilot of the helicopter turned his head towards the team and wished them luck with a mysterious smile in his corner of his lips. The helicopter dropped the team members one by one.

Now the task was to construct the target for which the team had to camp in the island and they were a bit apprehensive about the tracks. During the very first day they spent, the mystery of those tracks was unveiled in the evening. It was discovered that the tracks neither made by a truck nor any dinosaur rather it was the foot-prints of sea turtles. One of our team members saw a sea turtle moving he immediately called others and the team instantly did burst into laughter. They finally got the answer to the mysterious smile by the helicopter pilot. Later, they came to know that it was very common in those far flung islands that the tracks made by turtles were misunderstood as tracks made by trucks or tanks.

The Intense Labour

Col Misra and Mr Rane inside the temporary shelter

With our team on the ground, we were on board INS Rajput and were systematically preparing ourselves for the launch. The team was dropped on the island with all materials required for completing the target construction. No information was available regarding the rise and fall of tides. Still, they ventured onto that unknown island. We were unaware of the amount of work that had to be carried out.

The team made temporary shelter for themselves. But, as the time progressed, heavy winds started to blow during the dark which was later accompanied by rain. The intensity was such that it blew away our shelter, i.e., our tents, along with it. It kept raining for the next 24 hours and our team spent one of the worst parts of time over there. So how did the team survived? They wrapped the tent material around them and sat on the sand with the lantern as the only ray of hope until next morning. The team members passed that tough time by sharing their lifetime experiences.

In addition to their surprise, they had to face another consequence of being on the island – the Crabs. A large number of crabs attacked the edibles, cigarettes and anything they left over the ground. The most needed sleep was refrained from the crew as there was a probable chance of being attacked by crabs. So they collected dry coconut leaves around them and lit fire to keep away the crabs. Turn by turn, each one of them kept awake to let others take some rest.

After around 24 hours, the rain slowly ceased and now the sun took over the role of rain. The five member team started taking various measurements to construct the target which took one full day. As there were some rocky barriers, the process of aligning the target got delayed. Finally, the place was adjusted and the construction started.

Here, I have to mention the support that was extended by the Combined Command of the Andamans. The command deployed one helicopter exclusively for this flight trial for ferrying between the island and the base. If I ask our team members about the best breakfast they had in their entire life, without giving a second thought they would say that the one they had on the Trak Island. Upon being asked, one of the crew members said, "After being soaked in rain for more than 24 hours, the Aloo Parantha and the Hot Tea brought and dropped by the helicopter crew gave a new life to us."

The erection of the target structure was in progress. The target was divided into a lower section and an upper section. The lower section was a small pillar like structure upon which the upper section had to be placed. The placing of one section over another had to be done by a helicopter. This could be a perfect structure to simulate a shore based target for the missile. But how a structure such as this can stand still over sand? For that, they used pegs to secure the stand structure. The team had to spend the day without any shelter in the scorching sun due to which the team members got their skins to sunburn. Now, the upper section of the target had to arrive which was brought by the helicopter.

Col Misra at work with sledge hammer

The team stood underneath the chopper to place the upper

section on the tower. Have you ever noticed the gestures of people standing underneath a landing helicopter on a beach? If not, the experience may be well understood by the statement "It felt like thousands of pins pricking simultaneously all over the body", a team member upon being asked after finishing his task said. Finally, both the sections were placed as designated but again our fortune again tried to test our will. During the process, the upper section got placed over the tower in some angle which was not as per the plan. This meant that the upper section has to be turned for best fit which was a herculean task in itself. Military personnel are trained to think out of box for achieving crucial tasks when conventional ideas don't work. Col. Misra and Rane found out the solution by improvizing a turn buckle. They erected an extra peg on the ground and tied it with the iron cables attached to the structure through a turn buckle. Then using that turn buckle, they slowly turned the base structure to achieve the required angle.

Another issue was the video recording of the missile flight. The cameras were positioned at various points to cover the approaching missile. From where the cameras draw power? The team landed on that island with UPS systems which served as source of power supply to the cameras.

I went to Trak Island in a helicopter and requested the pilot to drop me on the island to meet the team. R. Adm. Nair, the Deputy Chief of the Combined Command who accompanied me refused to do so keeping in mind the safety aspects. He, knowing my adventurous mind, asked his men to tie up tightly with the seat belt. I asked the helicopter pilot to fly as low as possible close to the target. The helicopter made second round. I could see the target erected and I was satisfied. After all these actions, the team returned to INS Rajput after five days of hectic work on the island. They were tired, sick with full of blisters and pain. The missile was launched from INS Rajput. It flew all the way following the exact trajectory and penetrated the target. It was a perfect hit as confirmed by the helicopter crew. I also went to hover around the target and made sure that the target was hit. Our team got down on the island and collected the high speed video cameras. The videos confirmed

the precise hit on the target. The trial was a grand success! We have proved one more version of the missile - sea to land. What a great achievement and satisfaction to the BrahMos team who put their whole-hearted efforts to locate the island, get the ownership of the island for conducting the launch, construction of the target in odd conditions and physical exertion, carrying out the launch at the Andaman Sea far from our operating shores. It was a great effort, yet the mission got accomplished with perfection. There is no equivalent experience of this sort except that we have read about adventures of Sindbad.

When I met the members of the target erection team at Port Blair after the successful mission, my eye sight got blurred with tears. It was because of those men, they were badly affected by the harsh weather conditions with blisters all over the body, sunken eyes, and unshaven faces stuck with sand, but light glittering in their eyes. I can never forget that sight throughout my life. I have seen this light in the young scientists hardened by determination, working in the labs and missile integration buildings, without realizing the time - space boundaries. I have seen this light in the eyes of people who burst into tears after seeing through the missile meeting its mission objectives. Once I visited the frontiers, I saw the same light in the eyes of our soldiers who fought with the enemy disregarding the bad terrain, weather conditions, and other personal emotions. It is the light of determination, courage and dedication that keeps the nation on the track of development. What a great Nation we belong to and the men of action when there is need! Hats off to my friends! They are the heroes of BrahMos.

FIRST LAUNCH FROM INS RAJPUT - CLENCHED FISTS

I have already described, about the intensive cooperation we have with Indian Navy for the flight testing, induction and deployment of BRAHMOS missiles. It was Navy's foresight to possess such a powerful weapon onboard their warships. Upon seeing the first flight trial, V.Adm. John DeSilva, the then FOC-in-C of Eastern Naval Command thought of finding an opportunity to get a ship for fitment of BRAHMOS. Fortunately, INS Rajput was on re-fitment. He advised Adm Sushil Kumar, the then Chief of Naval Staff to give this opportunity to BrahMos. When Admiral asked me over telephone, I immediately accepted and gave an accelerated schedule to match the refit time and install BRAHMOS. It was a great boon for us.

Integrating the missile with a vessel means that the missile should be connected electronically, physically and mechanically, and it takes huge amount of time. Electronically, the missile can be integrated by first installing the Fire Control System (FCS). Then the FCS is integrated with the shipboard electronic consoles. Similarly, the launcher gets integrated with the vessel. Our design team interacted closely with the Russian specialists who originally designed the R-Class ships to get the structural and load details and worked out the details of removing the existing launchers, reinforcement required and the new launchers to be fitted. Preliminary project report was prepared and reviewed with the design review committee involving Indian and Russian Specialists, members from Larsen & Toubro (L&T) and Naval Dockyard, and officers from Eastern Naval Command. A comprehensive approach was worked out for fabrication and fitment of the structure, launcher and also the electrical interface requirements for the FCS. The work progressed at the Naval Dockyard and L&T. All the tasks were carried out as per schedule within the re-fitment time.

Then it was a time for a flight trial to validate the performance of the missile and ship complex specifically made for BRAHMOS. In that trial the missile had to be launched while the ship was moving. The whole mission was reviewed thoroughly by the appointed Flight Readiness Review (FRR) team. A number of issues came up like the adequacy of the software in the FCS, the trajectory of the missile at take-off when the ship is pitching and rolling amidst extreme sea conditions, the possibility of the nose cap falling on the deck, the effect of the full blast of the booster on the deck at take-off, the type of sound which the crew had to handle, safety of the people, limits on the speed of the ship and so on. It was the first time, a missile of this size being launched by the Navy and the concerns were well understood. Risk is involved everywhere, but we were not afraid of the risks. A careful intelligent analysis is necessary to overcome any risk. With this determination, we cleared the launch. The launch was a great success. Now, Navy got a big boon that a long range supersonic anti-ship cruise missile was inducted onboard the flag ship of INS Rajput - A historical moment for India.

Sinking a Ship

The Indian Navy decided to fire BRAHMOS from INS Rajput for various flight profiles. So, as a next step the missile's targeting capability had to be validated. For every flight test we need a target. Generally, the procedure is to use decommissioned vessels as targets. Since, they are actual ships they replicate the target parameters accurately for any missile. So, navies generally reserve the decommissioned vessels to be used as targets in future. For this mission, Indian Navy gave one of their decommissioned vessels Ex INS Androth as a target. Being a decommissioned vessel, the engine, weapons and all other useful sections had been stripped off the vessel, and it had to be towed to the designated place for the trial. But the vessel was in such a bad condition that after reaching the chosen place water started pouring inside. So, the flight trial became a question mark.

Fortunately, Commodore Chinnaverriya at BrahMos who had formerly been the captain of that very same vessel was in the ship which towed the target ship. Generally, the captains of Ships consider their vessels as their close companions. So, our man said, "I am not going to leave my companion sink" and he asked for volunteers. Men from the navy stepped forward who once were under his command. A small team was thus assembled. The team went to the vessel in a boat fighting the rough waves in sheer darkness. They worked all night to pump out the water. The vessel was afloat by daybreak and was staged at the designated place early in the morning. The target was declared ready. The contribution made by Cmde. Chinnaverriya and his team saved the mission.

Mr. Venugopalan, Mr. Kieslev and I reviewed the readiness and cleared the launch. The missile was launched in the morning on dot as declared. It took off beautifully with a roaring sound and pierced the target - Ex INS Androth. A helicopter was sent to investigate about the target's status. Commander SM Shiva Kumar, who was onboard the Helicopter, was continuously updating me on whatever he could see. He said "Target is located but it looks normal". As the helicopter approached the vessel, he reported a "hole in the target ship". The Helicopter was taken around the other side of the target ship after which he reported a "big hole in the target". That means the missile had pierced through target ship one metre above the water level, exactly hitting the middle of the hull. This is called Bull's eye hit. The trial was a great success and the Navy celebrated a grand party at sea onboard INS Rajput. What a grand scene of joy it was to see the jubilant team members!!

Lethality from the Speed

Combat Launch - One Shot One Kill

Having seen the successive successful flight trials, with full user participation, a request came from Adm. Arun Prakash who had then taken over as the Chief of Naval Staff, to conduct a combat flight trial during their annual naval exercise. Normally, trials do not include testing with warhead. With the assurance of giving production orders and no more trials to be conducted, he demanded successful destruction of the target by BRAHMOS. I accepted the demand though it was a deviation from our plan. Our Russian friends said that this test must be conducted by the Navy on a missile to be declared as delivered. Navy accepted that. We quickly mobilized ourselves to conduct a test with live warhead on combat mode, off the Goa coast. INS Rajput was loaded with a missile and Ex INS Sindhudurg was chosen as the target ship. On 15th April 2005, I was onboard INS Virat along with Adm. Arun Prakash, V.Adm. Sureesh Mehta, the then DCNS and other officers from Navy. Our team with Venugopalan were onboard INS Rajput. We carried out the launch with a precise hit on the target which split into two parts and sunk. All happened in no time. After the launch, we boarded a helicopter and went to see the condition of the target. When we reached there, vessel had sunk completely leaving behind no signs of its presence and only traces of the vessel's debris found. We saw a boat nearby from which naval divers waving their hands.

Naval divers had recorded the video of the missile hitting the target with a big bang, the target ship got split into two and sunk in waters. Later they said that even though they recorded the event, it happened so fast that the missile came and hit the vessel and it sank within minutes. The disturbance caused due to the sinking of vessel in water made things bit tougher for those divers in maintaining stability of their boat. The videos displayed enormous kinetic energy and the power of the missile. There is no match to BRAHMOS. The whole world came to know what BRAHMOS really means. Thereafter, there was no question from Indian Navy. Adm. Arun Prakash kept his words and placed production orders to fit BRAHMOS on multiple ships. He also gave funds to design and develop the Universal Vertical Launcher Module for fitment in the future ships. Indeed, it was a great partnership with Indian Navy.

DESERT LAUNCH – CONVERTING ANTI-SHIP TO LAND ATTACK MISSILE

BRAHMOS was originally designed to perform as an anti-ship missile with a radar seeker. We have successfully proven the fact and got the missile inducted in the Indian Navy. In order to bring a new version for the Army, it had to be used as a land to land missile. In other words, the missile could be launched from a Mobile launcher equipped with command and control system against a target on land. The typical targets are ammunition dumps, missile bases, missile or ammunition factories, railway yards, airports, communication centres, critical bridges and so on. A ship on the sea can easily be homed on by the radar seeker. But, for the Army targets, a suitable seeker is required for homing onto these types of land targets. Generally, the option is to go for Terrain mapping and GPS guided system. We started developing Scene Correlation and Area Navigation (SCAN) seeker. But, to quickly work around a solution for the Army, we had to adopt innovative methods which nobody attempted ever.

Testing the Seeker

Generally, a seeker fitted over an aircraft mimics the missile trajectory (but not the speed). The output of the seeker is analysed and from there conclusions can be arrived upon how well a target is detected, which trajectory it followed, what has been the angle etc. So an arrangement was made to conduct this analysis on a suitably modified Russian Tupolev-134 short range transport aircraft. We thought of flying the aircraft specially fitted with the missile's seeker over various real targets to study their signatures in order to establish the adequacy of the seeker on land attack mode. Our team reviewed the provisions of the telemetry systems and were happy with all the arrangements used to monitor the recording of seeker on targets. Our scientists were waiting for the Russian team, to carry out the experiments.

But that aircraft came with a string attached. It was conveyed that Indians were not permitted to fly in that aircraft as the aircraft belonged to another Russian agency, over which our Partner had no control. We strongly objected to it. We intimated that since it is a JV, Indian scientists have equal responsibility and right to conduct the experiment over our land. We intimated the Russian agency to take back their aircraft as we were not in a mood to carry out any test over our land without Indians. The team accompanying the aircraft understood the stubborn stand of the Indian side and contacted their authorities. Because of our stand, they relented and we could utilize the aircraft. We guided the aircraft to fly over variety of targets thereby evaluating the performance of seeker. We got excellent results and cleared the seeker for integration with the missile.

Flight Trial at Desert

We conducted two consecutive successful trials for proving the land attack version of BRAHMOS. These trials are carried out at the Pokhran Field Firing Range (PFFR). Both the flights were witnessed by Gen. J J Singh, the then Chief of Army Staff. Incidentally, the second trial conducted on 31 May 2006 became the summer trial. Rajasthan deserts are known for their heat, and in summers the temperature would easily exceed 52°C. The conditions were such that the work had to be started very early in the morning and was stopped by 10 AM. Again after 5 PM work started. Here, I must mention about the will power of our Russian partners. When we declared for the launch, they wanted to participate even though the temperature did not suit them at all. In their country -20°C is a normal ambient temperature. They were enduring the extremes of temperature but the sprit showed by them was tremendous.

Building a Target

The flight objective was to launch the missile outside the PFFR, flying over two villages travelling on the desert and hit a concrete wall specifically erected for this purpose at the designated target point. Constructing a wall in the hot Rajasthan desert was the most difficult part since the wall could not stand without a foundation which can only be constructed by digging the sand. So, a concrete wall was constructed which would lean over a supporting pillar like a structure analogous to the photograph stand.

Target Hit

Gen. J J Singh reviewed the preparedness for the launch. He went through the Mobile Command Post, Mobile Autonomous Launcher loaded with the missile, its standby, and the command centre. I gave a presentation to him on the mission objectives and the results of the FRR committee. The count-down was resumed and as usual at T_0 the missile took off beautifully, vertically up, turned towards the direction of the target and quickly disappeared from our sight. The General and I quickly rushed to a helicopter to see the effect on the target. We found that the missile had finished its job long back. We built a helipad near the target for the General. From the helipad, we went by a four wheel drive on the desert and looked at the target. A glowing smile came from the General's face. He congratulated me and said that the induction process must

go on a fast track for the Artillery division of the Army. BRAHMOS brought Customers delight. We got production order for one regiment from Army.

Superstition or God's Presence!

When we celebrate success, we must remember how our young team members went through the difficult experience. Two of our engineers who were in their mid twenties were present in a telemetry vehicle five to six kilometres behind the target. The cables of high speed cameras located near the target to videograph the hitting sequence were connected to the telemetry vehicles. The countdown was already initiated. It was just 90 minutes to time zero. Suddenly, the telemetry vehicle lost its signals. The two young engineers went in a vehicle to inspect the cable as that would have been the only possibility. They connected the cable which got snapped due to wind but while returning the vehicle got stuck in the sand somewhere near the target. Unfortunately, their cellular phones were not reachable and by this time the countdown had reached 45 minutes to zero.

There were speculations that the vehicle may produce some radar signatures providing a clutter for the missile. They were not ready to take chances and made clear to halt the countdown. So they started running in the desert towards the telemetry vehicle by following the tracks of the vehicle. Those who had walked in the sand dunes of a desert under the hot sun can understand how tough it would have been. They ran with their feet burying in the sand and reached the telemetry vehicle and asked an army officer to intimate the problem in the wireless radio. They felt tired. The army jawans who were patrolling the area saw these engineers and helped them. Colonel Misra who was associated with our development activities rushed to his wireless radio to contact the launch point for holding the countdown. He was planning to hold the launch for some 30 minutes so that he can send a recovery vehicle to recover the stuck vehicle. But unfortunately he was unable to establish contact and now the time left was just 7 minutes to launch. He rushed to the spot in his vehicle and commanded the driver to turn back the recovery vehicle. From that point, they could see the target and after moving some 500 metres away from the target, they heard sonic boom and banging noise simultaneously. The missile had penetrated the target, further travelled, and made its impact on the ground.

We were waiting at the launch point and could not establish communication from the lost telemetry vehicle. But, from other vehicles we got the information that the target was hit. Finally, the radio crackled and the Colonel gave the information; we went by an army helicopter and saw the damage on the target.

But in private he told me that the two engineers were rushed to the nearby military medical facility. I also rushed along with the senior scientists to medical facility. When they saw me, they asked a question, "Sir, is it a success?" I was amazed to see the commitment of the young scientists irrespective of their difficulties. With tears in my eyes, I nodded my head in affirmation and blessed them. The grit and dedication of these boys moved me a lot. This reinforced my thoughts that a stronger new generation is getting ready to see the uplift of the nation.

TASTE OF A FAILURE AND REGROUPING

After successfully inducting the First Regiment from 2007, the demand started coming from Army for a next version of BRAHMOS which could discriminate a particular target of interest from among a group of targets. This needs surgical precision to locate and hit the target. More intelligence needs to be incorporated in the missile onboard computer and guidance system. We required a GPS receiver which would be compatible with multiple Global Navigation Satellite Systems and a seeker tuned to home on non-radio contrast targets. This requirement was quite challenging and we started developing the system as BRAHMOS Block-II. The first launch of Block-II was scheduled on 20th January 2009. General Deepak Kapoor, the then Chief of Army Staff came to witness.

In the Pokhran range, five different targets were constructed. One of the targets was not reflecting the radar energy (low signature) as it was constructed with a thin sheet. This was chosen as the target. To simulate the conditions of urban terrain two false targets were erected along the sides of the actual target forcing the missile to identify the target from the two false targets. The trial was conducted on the set day.

The missile took off, discriminated other ones and headed towards the actual target. To our disappointment, the missile did not hit the target instead it just passed above it. Gen. Kapoor was dissatisfied with the performance and the Army informed the media that the flight trial was a failure. The press took it very seriously because in the history of BrahMos, this was the first incident of missile not hitting the target. Moreover, it was a crucial time for our country since it was immediately after the 26/11 Mumbai attack. A huge outcry was created and the Army maintained an indifferent posture by saying that further induction will be carried out only after successful flight trials. I assured the Govt and the Press that we will find a solution to the problem. Within a month we have found a software glitch in the algorithm of Block-II missile.

Courage to Succeed

When I returned to New Delhi I saw pale faces in the office. For the first time, my team tasted failure which moved them so much that they were unable to utter the word hello! Unable to bear the brunt of the failure and more than that the criticizing comments had hurt the team. I called for a meeting in the evening where I told, "We will conduct a successful flight trial in a month after re-tuning the software."

I already saw a major setback when I was at ISRO, during our first attempt to launch SLV-3 as the vehicle failed in its mission on 10th August 1979. The same thing happened in the fourth launch of Prithvi in 1989 and second launch of Agni in 1991. Failures are grounds for a lot of learning if we carry out the analysis carefully and draw the conclusions correctly. The second launch of SLV-3 was a great success and further flights of Prithvi and Agni were consecutive successes. I narrated my experience of Agni to my team.

Agni Experience:

You may recall how the REX Mission thought by Dr. Kalam became Agni in IGMDP as brought out in Chapter 3. Initially, Agni was a technology demonstrator to establish the re-entry technology with a two stage propulsion system and the carbon composite re-entry structure with guidance and control. The first stage was a solid booster which was used in the first stage of SLV-3, with its control system: SITVC (Secondary Injection Thrust Vector Control) and FTC (Fin Tip Control). We got the first stage from ISRO and went for a launch in 1989.

The Agni launch campaign was a special one, full of technical, managerial, social and political challenges. Agni was all set to make its first test flight in April 1989 after six years of laborious effort in the development of various subsystems and large number of ground tests. Launch campaign commenced on 31 March 1989. A special launch complex was made operational in the Missile Test Range at Chandipur with many instruments for tracking and telemetry. The mission control centre links range safety, Block House, SHAR Range, down range stations and naval ships. The launch countdown started on April 20, 1989. At T–6 minutes, the count-down went into the automatic mode controlled by the check out system. The missile was put in its internal power through the actuation of batteries and the pyros fired. All the systems got actuated and checks were continuously carried out one by one to keep the missile ready for lift off. At T–1 second the launch was aborted due to hold given by the computer as there was a problem in the umbilical pull out. The hold was not lifted by the Mission Director in time resulting aborting of the mission. It required a detailed review of all the power packages before the authorization of launch. There were many important visitors—the Defence Minister, Cabinet Secretary, Defence Secretary, three Chiefs of Staff, and many others. Everyone was disappointed to see the missile not taking off. Before chaos could take over, Dr. Kalam took control over the situation. Detailed analysis of the failure was carried out on the same day and rectification started. The whole system got refurbished and special effort was made to clean up the first stage control system with the help of ISRO team. The system was made ready once again within 10 days and launch was attempted on May 1, 1989. Again, during the countdown at T–3 seconds there was a failure in the first stage control system as one of the SITVC injection valves gave way leading to leakage of the secondary injectant. The mission was aborted by the automatic check out system. There was a crisis. The Government was very unhappy because of the failure of the missile to take off in two attempts. There was a crisis in DRDO and de-motivation as the Agni launch was put on hold by the Government. The newspapers of those days were full of cartoons, depicting Agni sarcastically as IDBM (Interminably Delayed Ballistic Missile). The situation was grim. We did a quick failure analysis of the control system package and found out the proportional injection valve cracked due to stress corrosion, induced by the residual Strontium Perchlorate. Many components in the control system of first stage had to be redesigned and fabricated. The system mounted on the Base Shroud had to be integrated and tested to environmental requirements before it can be cleared for use in the missile.

I was tasked to rush to ISRO, Thiruvananthapuram, some 2000 km away from the launch pad carrying the vital control system package of Agni. The team worked round the clock for 12 days. Detailed analysis of the component failure and rectification action for the first stage control system was completed in record time. The failure was analyzed thoroughly; design modifications were incorporated in the components and assembly, new components fabricated and the packages integrated in a clean room facility. The integrated system went through all acceptance tests and the system was cleared by a specialist committee. As I had worked in ISRO with the same team for SLV-3, it became easier for me to realize all these tasks in 12 days which normally would have taken 60–70 days. No other go; we had to work to make Agni mission a success; the team started working 24 hours daily without breaks. We neither cared about our dresses, our appearances nor food and sleep; only the mission was in our mind. This was because of the tremendous support of technologists and the urge of scientists to ensure success. It was really amazing to observe how hundreds of scientists and staff members of ISRO could work continuously and ensured the system readiness with all acceptance tests in 12 days time.

The new system was flown to the launch pad in the same aircraft which was waiting all the 12 days at Trivandrum Airport. We landed at Hyderabad to present the test results to the Flight Readiness Committee. The systems were flown to the Range. When I reached DRDL and met Dr. Kalam, he informed me that the Government has not cleared the launch in spite of our best efforts. With complete information assuring the successful launch, we made an attempt to convince the concerned authorities at Delhi to believe in technologists and clear the launch. We succeeded. Launch campaign started once again in the midst of continuous criticism from the press and other external pressures. The final launch date was dictated by the hostile weather conditions including a threatening cyclone. The Meteorological Department and all partner institutions worked together as the ambient conditions created great anxiety. All work centres were connected through Satcom and HF links. For the entire campaign, Indian Airlines and Indian Railways with just one request had arranged movements of hundreds of the members for the launch campaign without any advance reservations. The Air Force came to the rescue with its continuous support by providing aircraft and helicopters. The Navy through the ship borne stations even in bad weather conditions and Army through its medical facilities also helped. At Chandipur at sea, about 12,000 people had to be evacuated for safety reasons, from an area of about 3.5 km radius from the launch pad. A combined effort was made with the local administration to move people, give them proper shelter and food and make necessary arrangements for their safety on the launch day. Finally the day came. On 22 May 1989, the count-down was smooth and Agni took off majestically and completed its mission successfully. The criticisms, abuses, and de-motivating actions for delay in launch now turned into praise and the whole country rejoiced in success. It was a great day for India as we entered into the Elite Club having long range missiles. To tell you further, on 24th May 1989, a cyclone hit the range and heavy rain started pouring in the following days.

I told my team members, **"the factor between us and the success is just a matter of time; the challenge is to compress the time."** I asked them, "If I was instrumental to do it 20 years before, why cannot you people do it now?" I got an affirmative answer as expected, because I was able to see the faces slowly getting rid of the bitterness of failure when I started narrating the Agni experience. The entire team geared up and everybody regardless of their work profile contributed.

Target hit on 29th March 2009

While upgrading the missile, even though we would have done all possible simulations, there could still be probabilities of failure due to unknown reason. So, a detailed analysis was carried out with all the concerned engineers and a programme was chalked out along with

Russian Specialists to validate the new software and to carry out necessary simulations.

The corrective measures included more sensitivity to the seeker to track low radar energy from the target with a dynamic computations to give error details to onboard computer and logic for correction with the INS. All these problems do not occur with long range mission which is the real application. Here, in the desert, the range is very limited and the time of flight is short and the speed is supersonic. So the software corrections were incorporated taking into account these measures and the whole mission was simulated for a number of times. This time we were quite sure that the target will be hit. Hence, we informed the Army that we were ready. There was a lot of hesitance in participation for the launch. Obviously, new pressure came for looking at alternate imported systems. Disregarding all these apprehensions, we proceeded for the desert launch with great confidence. The launch took place on 04 March 2009 and the missile hit the target. This time, the Deputy Chief of Army Staff came to witness the trial. We declared success. Army said that they wanted one more launch to repeat successful mission to establish the reliability of the missile. I smiled and immediately agreed and said "it will be done." Multiple targets were created in the deserts out of which the army wanted a particular target to be identified and hit by the missile. The target chosen was the least signal emitting thin sheet surrounded by the targets giving powerful signature. The missile was to hit only the least signal emitting target. This was a real challenge. On 29th March 2009, the flight trial was conducted, which was witnessed by the Director General of Military Operations. The missile discriminated the specific target among the group of targets, flew straight towards it and hit precisely. This was a grand success with a bull's eye hit on the target. We declared the Block-II readiness to all. Army men who witnessed the flight got thrilled and the Army Headquarters applauded. With that result, we got production order for two more Regiments of the Army.

Failure is an inherent part of any high technology development, particularly, if we try to achieve new missions stretching the potential of the weapon. We cannot avoid it but technologists know that even failures give better understanding of the system and learning for the future. But for the user who wants the latest weapon to be with him has no time to loose. Therefore, he cannot tolerate a failure even though it is first time attempted. The above problem came in the first flight of Block-II and we could quickly recover from it and show successive successful launches.

The media celebrated the success of the Block-II and the way the failure was managed with guts. Sujan Dutta and G.S. Mudur wrote in The Telegraph on April 03, 2009, about the success of BRAHMOS.

"Missile Man pierces army apathy

How a tenacious technocrat put the supersonic BrahMos back on track

New Delhi, April 3: India can stake claim to be among the first in the world to be ready with a supersonic land-attack cruise missile because of the tenacity of an unheralded Missile Man whose pet project was almost written off for aiming too high.

Sivathanu Pillai, a technocrat whose bald pate is not covered by berets, whose chest is bereft of medals and shoulders of epaulettes, dared the Indian Army by claiming he would arm its artillery divisions with a missile the world had not seen.

........"

Concluding Remarks

I narrated a few of the typical experiences that BrahMos Team has gone through. There were many. On all occasions, Team BrahMos, both men and women have displayed high degree of integrity, commitment, courage and value during the last fifteen years. I was fortunate to have such a wonderful team with me. This was acknowledged by the great men who participated in the celebration that marked the completion of fifteen years of BrahMos as Aardhik Diwas (Partnership Day) on 19th February 2013. I will cover more on the views of great men in the later part of the book.

PART 6

All Hands on Deck

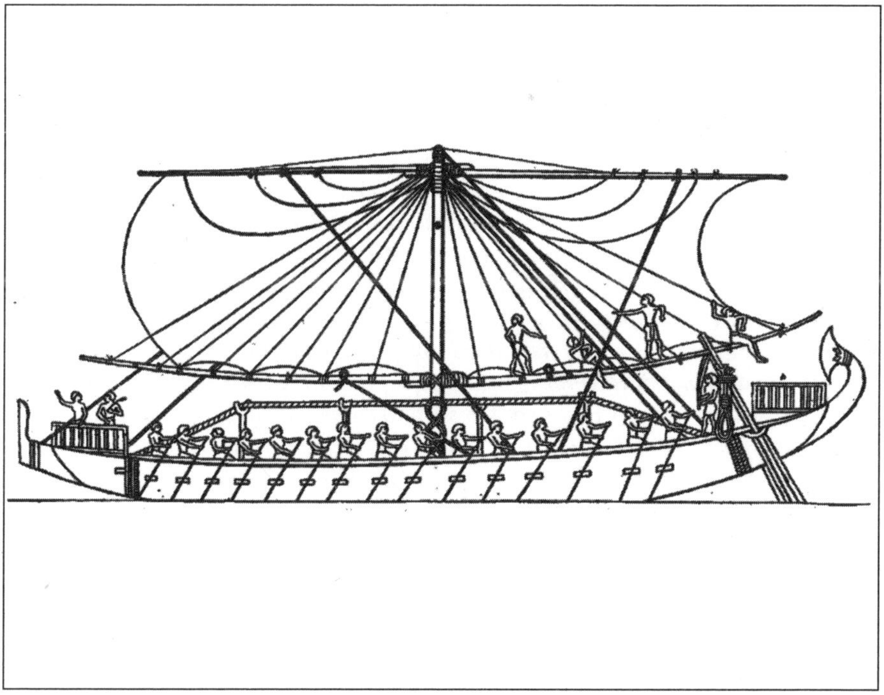

"Coming together is a beginning, keeping together is progress, and working together is success."

- Henry Ford

'All hands on the deck' is a naval terminology which was followed when the sail ships were in use. During rough weather or during an emergency, the ship has to be manned to the fullest efficiency. For such a situation, the captain instructs to raise a call 'All hands on the deck.' When this call is raised all the crew members and sometimes passengers too will rush to the deck to perform emergency actions like pumping water out of the ship, furling the sails etc. Only such a collective action will help the ship to come out of the crisis. Here in this part, I am going to tell you that how so many entities in India joined hands to establish the production lines for producing BRAHMOS in India. Only the collective effort yielded the result...

14

FUSION OF GREAT MINDS

Anybody will easily say that India and Russia are entirely different in culture, language, weather and so on. In due course of the project, we felt the enormity of those differences that served very well as barriers. Both the sides really put-in efforts to jump over these barriers to understand and co work with each other. Only our mutual determination led us to win over those barriers. I am narrating the effects of differences in working with culture, language, standards & specifications, climate & cuisine, the most intriguing barriers those existed in countless dimensions.

WORKING CULTURE

First in terms of working culture India is the South Pole and Russia is the North Pole. Working culture involves adherence of time and quality for work. According to the Russian working culture, both of these parameters are important stressing a work done with fullest quality and in the said time. For example Russian friends used to stick to office timings. They used to enter the office two minutes before the office opening time and will nowhere be seen in the office two minutes after the closing time. In contrast, the momentum increased unimaginably for Indians after the closing time of the office. It is the IGMDP practice where all the important issues were used to be discussed after 6 PM. We would go to bed very late after midnight and be back to the office on time. The office entry timings were slightly relaxed whereas leaving time would always be uncertain. One day, during IGMDP, a good old-friend came to meet me during one afternoon while I was in a review meeting. My assistant told him to come by 10 AM the next morning so that he could easily meet me. But he insisted in meeting me on that day itself. Again my assistant told that I might return late at night. My friend, who misunderstood that late meant somewhere around 8 PM, started waiting on a couch. When the meeting got over at 2 AM, I saw somebody was sleeping on a couch and it was none other than my old friend. He was in a deep slumber and was across seven seas and seven mountains. Russians can never accept this and will leave by 5.02 PM. In another incident, I was hurriedly returning to Vasant Vihar office one evening. I asked the security guard what the time was. He replied in Hindi, *"Saab abhi to gaadi Russian ko leke nikal chuki hai to isliyae time to panch panch hoga"*. He meant that the vehicle with Russians has left just now so the time should be 5.05 PM. It shows their promptness to keep up the office time.

Such were our working cultures; where for an Indian a quality work can be completed with a marginal relaxation in time whereas for Russians quality work has to be completed at punctual time. Russians even include that punctuality in the quality of work and many of our Russian friends used to say that a work not delivered at right time is inferior in quality. Ultimately we learnt that important lesson and started to follow the same. To our customers, our products will be delivered well ahead of time which is contrary to the Indian scenario. We had initially tough times in managing an acceptable time and slowly we got used to the Russian punctuality. Somewhere during the year 2003, a Russian specialist visited us. He got terrified that we used to sit up to 2 AM or 3 AM. Slowly he said, "Now I understand how you Indians developed missiles so quickly!" Indeed it was a revelation to us also! When I look back, I am able to see that we had been working approximately around 15-18 hours a day. You know that the regular activities enter your brain and in due course brain executes those activities easily and that we call it as habit. Some of the important decisions were only taken during midnights. Sitting late and continuously doing work became our habit. Gradually, our Russian colleagues also got accustomed to our late sittings.

LANGUAGE

The biggest barrier was language. Russia is almost a single race country with one language. On the other hand, India is a sub-continent full of different languages, their dialects and accents. So, in Indian national projects, the common connecting language has become English since India is the world's second largest English speaking population. All the Russian documents, names of the systems, their sub-sections used to be in Ruski (Russian language). Our documentations were all in English which Russians were not able to understand. So we were in need of interpreters and now emerged a funny situation.

The majority of the interpreters and translators were from arts and classics background. They had given up science in their high school levels itself. Unfortunately, what they had to translate was totally high-end technologies. So they were forced to learn the scientific jargons in both English and Ruski. Luckily, some of our scientists studied in Russia. They came in very handy during the meetings for interpretation. Though their Ruski wasn't that perfect, still they were capable of making the Russians understand our point and their point to us. Many of our scientists were able to learn Ruski in the course of time and Russians also learnt to speak English.

BRIDGING STANDARDS AND SPECIFICATIONS

Indian scientists are familiar with western standards in every aspect of Engineering from design to documentation, etc. But in Soviet Union, which was once known as the iron curtain, with everything shrouded with secrecy, followed their own set of standards and specifications called as GOST. Say for example, a bolt of Soviet origin would never fit with another western country's nut and vice versa. Starting from this, miniscule pieces to huge and complex systems, there ought to be mismatch. In places, wherever possible, the existent missile systems were matched with the Indian standards and under impossible conditions; we simply accepted the Russian standards for smooth functioning. This proved to be a strategic move as it reduced the time considerably for working with the Russians but it came with a price - 'difficulty.'

The electronic systems that have been developed in India were with Indian standards of protocols. This necessitated the development of new suitable intermediate systems that could match both the Russian and Indian protocols. These systems accepted Russian protocols and extracted data from them and sent out the data in the Indian standards of protocols.

Numbers were extensively used to denote the projects, developed systems, sections and subsections. This was very quite different from the naming conventions followed in India. Initially, the use of so many combinations of numbers used to confuse us a lot but we gradually accustomed to it. For instance, three Talwar class frigates were being built by Yantar shipyard at Kaliningrad in Russia for the Indian Navy, on which BRAHMOS missiles had to be installed as the primary weapon system. This means that the Indian design team and Industry staff had to work with the Russian designers very closely to understand each other. Especially, with respect to standards and operations, so that the ship and missile system become compatible to deliver the desired output.

CUISINE, CULTURE AND CLIMATE

Culturally both India and Russia are conservative countries but the way people get mixed with each other is different. During our very first set of visits to Russia we used to get puzzled whether we are being greeted with a smile or not. Because the Russian smile used to be lesser visible on a casual look, but what is for sure is that they would be greeting us. Regarding the Indian smile I would quote one of our Russian colleague's words, "In the beginning we thought Indians are rude people due to their way of smile but we were wrong. India is a country filled with smiles and everybody smiles; it is good to feel the warmth of the smiles and we have to learn that."

Food claims the primary importance of man. Our team realized this at Russia. As we grow up, we develop a certain taste with respect to the cuisines we eat. For a change, other cuisines may be enjoyed but not on a daily basis. Indian cuisines are centred on spices. On the contrary, Russians prefer food without spices and thus the very basic necessity was not being fulfilled for the teams of both sides. Whenever the Russians visited India, the food used to be too spicy for them and when we went there, the food was tasteless. So, for both the sides, a tour to the partner nation used to result in stomach problems due to difference in food preferences. Gradually, as the visits became more frequent, we established a Russian corner in the Delhi office cafeteria and an Indian corner at NPOM. Now, we were not just sharing the missile technology expertise but also sharing the art of cookery. Russian ladies got specially trained to make Indian food at the NPOM kitchens.

The climatic conditions also pose severe problems to our scientists. When the preliminary design review was scheduled, it was winter season at its peak in Moscow. We were advised to buy some Russian winter clothing before setting off. But when we wore those warm clothes it was felt that our body weights had doubled. Without knowing what is going to happen we were joking that we had become astronauts, medieval knights, etc. When we reached Moscow it was severe cold. We were shivering even inside that armour like clothing. Somebody informed that the temperature was bit on the hotter side on that day. We were unable to control our smiles as the temperature was 25°C below zero. When exposed to such cold temperature, it felt as if there is no nose, no ears due to the numbness. We spent around three months over there and after returning back and shedding those special clothes, we felt weightless. Still today we sweat when we see or think of those clothes.

The Indian weather also troubled our Russian friends. One of our flight-trials was conducted in Pokhran during the last week of May. The ambient temperature was 52°C and the launch campaign was scheduled for 20 days. That test plunged the Russian friends who came to participate in the launch campaign. They wrapped themselves in a cloth soaked in water and also tried to find out the air conditioned shelters. They had a hard time in the desert as well as at Chandipur at sea due to high humidity. But on a mission oriented projects such barriers do exist and have to be crossed at any cost. We used to adjust ourselves according to the situation. After all, mission is bigger than anything else.

15

MEN OF WISDOM

During the course of our journey we came across many people. Some good, some bad, some supportive while some menacing. We had to interact, work with many top politicians, ministers, national leaders, service chiefs, senior scientists, bureaucrats, etc. of our and other nationalities. We had many nice experiences and some experiences that will ever cherish in our memories. I am narrating some our good memories here.

THE FIRST FLIGHT TRIAL

The design of the missile was decided by the time the company became operational. Initially, we were lacking with facilities to integrate the missile in India since DRDO was not dealing with such kind of missiles. So all the necessary components from DRDO were sent to Russia and they were integrated with the remaining Russian components. The final missile was secretly brought by the Indian Air Force transport aircraft. Even the transport crew remained clueless about the content of the container. At that time, only the top brass of the three services, top bureaucrats of defence ministry and DRDO knew about the project. A date was fixed for the trial as 12th June 2001. 12th June is celebrated in Russia as Russia day.

We were pretty confident and at the same time very much nervous also. Simply saying all those who were involved in the process could feel butterflies in their stomach. Mr. Jaswant Singh (the then Defence Minister and Minister for External Affairs), the three service chiefs, top brass from DRDO and NPOM had assembled at Integrated Test Range (ITR), Odisha. Only after seeing such VIPs, the lower and middle level staff at the ITR felt that something important was about to happen in the range. The countdown began and everyone's attention was on the monitor in the blockhouse. At T_0, the missile came out of the launcher like a sword coming out of its sheath. "From that moment a new era in missile warfare began."

Since the project was a top-secret, initially there was a dilemma about releasing the news to the media or not. But the then Prime Minister Mr. Atal Bihari Vajpayee approved the release of news to media. When the news about the trial got released, the whole world was stunned.

From that day itself, the company started to grow exponentially and every month there has been steady progress. Before the flight trial, nothing was sure like whether the defence services would buy the weapon and in case if they don't what would be the future of the project, etc. But within minutes the missile cleansed all those negative thoughts and injected confidence to the users, particularly to the Navy.

INAUGURATION OF CORPORATE OFFICE BY VLADIMIR PUTIN

Almost a year after the 1st flight trial, we got a land for our corporate office in the New Delhi Cantonment area. Initially we were allotted a different land but that was not made available to us due to some issues. But the second time the land allotted was really huge and was highly suitable for our purpose. We decided to construct a guest house and a state-of-the-art corporate office over that land. We planned to get our corporate office inaugurated through Vladimir Vladimirovich Putin, the then President of the Russian Federation. In the year 2004, our office building was almost ready. We wanted the office to hold a permanent exhibition of the history of the Indo-Russian relationship and the technological prowess of NPOM and DRDO - the shareholders. This exhibition is an attraction amidst school children and it is open to students on Saturdays. We also host an exhibition of Dr. Kalam's photographs, laurels, awards, related articles, etc. All his awards including his Bharat Ratna are displayed in the exhibition. Also, he generously donated his very favourite Veena. Dr. Kalam is one of the founders and we were proud to host an exhibition of his articles.

As Mr. Putin's visit came closer and closer we came across some unforeseen problems in construction and found ourselves lagging behind the schedule. The security agencies visited the buildings and the surrounding areas for security purposes. That team consisted of some Russian protocol officials and after they surveyed the incomplete structures they started bawling. They conveyed to us that they could not include the inaugural function in the itinerary of the Russian President. Finally, we convinced them that at the time of the President's arrival everything would be in place. We were now left with only 10 days prior to the ceremony.

We had to apply concurrent engineering practices to carry out variety of work in rapid phases. Masonry work, electrical and plumbing fittings were carried out simultaneously along with interior decoration. One night before the ceremony, all the tasks were complete for the security agencies to take control of the buildings. Now, it was exactly 12 hours before the function.

The next day, Mr. Putin came and was very happy to associate himself with a world class high technology project resulting out of an international cooperation between India and Russia. During his speech, he lauded the successful joint venture.

Instead of staying with us for 20 minutes, unlike his itinerary, Mr. Putin spent more time at the complex and showed great interest in knowing the details. He also addressed this fact to the press as well.

He praised BrahMos as **"Indo-Russian cruise missile joint venture**

BrahMos is a brilliant example of pooling of global efforts in developing new weapon systems and it should be the model for all future joint ventures with foreign countries. BrahMos joint venture co-operation in a short time has traveled the path of designing experimental models to serial production and supplies for induction into services." Mr. Putin has been a staunch supporter for BrahMos project and whenever he gets a chance he mentions about BrahMos in his speeches.

In the year 2005, when I went to Kremlin with the then President of India, H.E Dr. APJ Abdul Kalam on an official visit, I had the opportunity to convey to President Putin that his recently issued decree makes the JV difficult to function as every interaction with Russian Industries had to go through Rosoboronexport. I requested for a waiver for the joint venture BrahMos, as there will be difficulties in the procedure leading to heavy delays. President Putin noted the point and soon a special decree of the President was issued maintaining the relationship between BrahMos and NPOM, giving a major boon to the BrahMos JV to progress on a faster track.

In another incident during MAKS 2011, Mr. Vladimir Putin, the then Prime Minister came to witness the Hypersonic programme of consortium Institutes of Russia. While entering the Hall, he located me standing away, rushed towards me along with the accompanying officials, hugged me and got apprised of the latest status of the JV. This simply shows how a leader of his stature gives importance to collaborative programmes where success is delivered.

FIRST DELIVERY OF THE LAND BASED SYSTEM

Dr. Kalam after becoming the president earned great fame as "The People's President." By the year 2007, our land attack version for the Army was ready and production process was progressing. We organised a ceremonial function on 21 June 2007 as a mark of the beginning of delivery of systems. It turned out to be a huge function attended by the Defence Minister Mr. A.K. Antony who declared that day marked a historic day for the Indian Army. Gen. JJ Singh who received the missile from the Supreme Commander of the Armed Forces expressed his happiness to posses the most potent missile for the Army. President Kalam declared that BrahMos has made India the first and has broken the Fifth Nation Syndrome.

INTERESTING BUSINESSES

The First Business

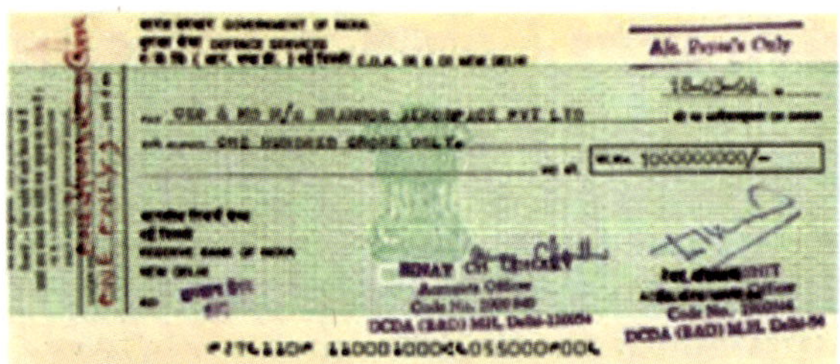

In the first flight trial, the missile was launched vertically And in the second trial it was launched in an inclined position. Immediately the then Chief of Indian Navy told that we should test the missile from their ship. Anyone would expect that the first business would have been on credit. No! We requested money from the Navy to integrate the missile in their ship. The chief looked at me stunned but sanctioned the money with a smile. This was the first money dealing with our customer and it so became that we started the business with a debt.

Business Expeditions

In military business sector, international military exhibitions play a vital role. These exhibitions are the launch pads for the business campaigns for selling military hardware. BrahMos Aerospace is an aggressive marketer in such military expos and it becomes necessary to fetch export orders. In the beginning, we wanted to establish and familiarize the name of the product internationally. In the year 2001, we participated in Langkawi International Maritime and Aerospace exhibition (LIMA). Though we were just a minor in the eyes of many, but our presence made to raise the eyebrows of many people. For other exhibitors and military business tycoons our presence was very strange. They were accustomed to the Indian presence as delegates and visitors only.

Initially, we were able to secure remote corners in the exhibitions hall due to our late registration. There was reluctance from all over in our company's participation in the expos but, after much debate, finally, the decision to participate was made. At the exhibition sites, we generally remained unnoticed and for those who could glance at us, reflected a feeling that they just saw an alien from Mars.

The then Prime Minster of Malaysia, Dr. Mahathir Mohamad, who was inaugurating the expo, came for a visit and with a curious look he entered our stall. There were only three of us; myself, the Indian Ambassador for Malaysia and Vice Admiral De Silva, the then Vice Chief of Indian Navy. From the outside, it was a small stall as we had few posters on the display and videos of the flight trials being played on the laptop. With these limited resources our three member team and especially the Vice Admiral did commendable presentation of the missile performance. After seeing the videos Dr. Mohamad felt impressed and instructed the Chiefs of the Defence Services to keep an eye on BRAHMOS as it could give an advantage with its accuracy and speed. In the meantime, other exhibitors became curious due to the amount of time spent by Dr. Mohamad at our stall. They gazed at us puzzled as if they were thinking what these guys would have done to engage the Prime Minister for so long. From that moment onwards, we were swarmed by various service personnel and other bureaucrats from the Defence Ministry of Malaysia.

Another interesting event was during "La Feria Internacional del Aire y del Espacio" or simply FIDAE. FIDAE is a defence expo conducted in Chile and an important event for the arms sellers. During this exhibition we were a well grown up company; a big team was representing BRAHMOS. Chile government enquired us about exporting the missile to them. Informally, officials revealed that Chile was going to buy the missile system from us. Later, Mr. Pranab Mukerjee, the then Defence Minister, was on an official visit to Chile. He conveyed the Chilean Defence Minister about BRAHMOS and ALH. Later, we came to know that someone blocked that deal. Marketing is not so easy, even with the best product we have. We have to manage the under currents also, with honesty. That is the challenge. When I visited Pentagon later, its Naval Fleet Commander had all the praise for BRAHMOS and was not ready to leave my hand. He also expressed that the Indian Navy possessed tremendous power with BRAHMOS as its prime strike weapon. An Indian Vice Admiral, who was with me, acknowledged the statement made by that Naval Fleet Commander.

In another international expo, a delegation from one of our neighbouring countries visited our stall. A Major General – the leader politely asked whether we can explain about the BRAHMOS system. As it was an exhibition, we were obliged to display the normal presentation material to highlight the importance of BRAHMOS and its capabilities. He expressed his happiness on the frankness and wished good relation between the two countries. He however whispered to me whether India is prepared to sell BRAHMOS to his country. I could not resist my temptation to respond to his question. I told him "BRAHMOS is a free delivery item to his country".

16

MISSILE INDUSTRY CONSORTIUM

So far I have explained how BRAHMOS is a unique missile jointly developed by India and Russia. At the same time, the major lacuna in indigenous defence production is the non-availability of quality products, produced in adequate numbers and realized in a predefined time. Moreover, the success of a Product depends on its competitive value in world scenario.

Product competitiveness is based on (a) Product Excellence (Performance, Quality & Reliability); (b) Cost-effectiveness and (c) Availability of required quantity in Time. Competitiveness in products leads to global leadership for the Nation. To achieve this unique position we require out of box concept, creative thinking, innovative approach and dynamic leadership. Above all, we need strong partnership with network of industries.

From the very beginning of the JV Programme, all our thoughts were to generate an industry consortium with Industries funding for themselves. Since, this being a JV, we could have utilized the large industry base available at Russia. This would have defeated the very basic purpose of JV as all the subsystems would have been produced only in Russia thereby increasing dependence on Russia for everything, which was undesirable. Therefore, a very articulated policy had to be evolved to build a missile industry consortium for BrahMos. While forming the JV, the Russians also appreciated that the missile should be produced in both the countries and together we must build the industrial capability to establish common standards, fabrication methods and inspection procedures in order to facilitate inter-changeability of systems. This would also facilitate the use of JV Product by Armed Forces of India as well as Russia.

At the Feasibility Study Stage, it was decided that the missile is going to be unique and the production orders will be large from many customers – India, Russia and friendly countries. Therefore, producing the missile in both the countries was an essential requirement - a perfect JV kind. Moreover, in case of any change in geopolitical situation, the missile could be produced in the respective country to meet the requirement and also continuous product support can be ensured. Therefore, consciously it was decided to approach the industries to become partners to the programme by making them to invest in the infrastructure.

I think, at this point, I should first narrate the present scenario of defence production in India. After the 1962 war, India needed production of defence equipment and weapons in large scale and resorted to technology transfers and production under license necessitating funding of infrastructure in the country. Therefore, the Government formed public sector undertakings and Ordnance Factories for production of defence equipment. There were no private industries that could come forward to establish facilities to compete with the PSUs. Over the years, the Private Industries started popping up but they continue to lose the battle as the Government gave PSUs preferential treatments in terms of choice as well as price. Since, they are owned by Government, there was no hesitance to invest in large scale and fix the contract as cost plus. Gradually, PSUs attained monopoly in industries relating to aircraft, electronics, vehicles, ship building, battle tanks, missiles and all the defence orders were placed with them. There was no chance for the private industries to grow. PSUs went for technology transfer and production for major systems like tanks, aircraft, missiles, radars, etc. They had to depend on foreign industries in terms of understanding the process and provision of spares. Over the years, this practice made the PSUs technologically inferior. Adding to this, they became larger and larger with huge manpower, out of proportion with respect to the productivity and finally resulted in large delays in the availing inefficient systems. The equipments became older and due to lack of modernization, these industries suffered technological obsolescence. As the Armed Forces required modern warfare, the foreign vendors started aggressive marketing which led to heavy import of equipment and weapons.

ISRO being an autonomous organization, started involving private industries and also utilizing the best from PSUs. Space vehicles needed special materials and chemicals, special fabrication technology, electronic systems which forced them to look for options resulting in birth of many private industries. The foresight of Space Commission especially from the visionary leadership of Prof. Satish

Dhawan motivated both public and private industries to show difference in performance. This culture saw some of the large scale industries like HAL, L&T, Godrej, Walchand Industries to specifically build Aerospace quality fabrication centres at their facilities. This transformed the outlook for private industries and changed the dependence only on PSUs. As Dr. Kalam and I were associated in building industrial base for launch vehicle in ISRO especially for SLV-3, the same approach was extended to IGMDP. Today, I must say that Private Industries have grown in stature in infrastructure, technological maturity and efficiency to undertake design, development and production of major defence equipment. It is time for a rational policy to utilize both public and private sectors in a judicial manner with an aim of quality production in large scale so that there is no dependence on imports. This means that the PSUs have to be made more efficient, productive and competitive since they have already invested and are investing huge funds from the Government. Appropriate restructuring is essential for these Defence PSUs to perform in order to ensure indigenous production at a higher self-reliance ratio. Moreover, private industries must come forward to have in-house R&D to support design and development, and also technology partnership at the frontline areas with other foreign collaborators. This will rejuvenate both public and private industries to contribute in large scale production of defence equipment within India.

THE BRAHMOS WAY

In the IGMDP era, many industries started building up core competencies in production of ferrous and non-ferrous alloys and composite materials, precision fabrication of components and large sized structures, electronic systems, avionics, sensors, radars, chemicals, cables, connectors, etc., from raw materials stage to systems. The industrial base therefore was strong enough for better utilization in the realization of BRAHMOS missile and ground systems. We needed to build around this capability and suggest augmentation of capacities to these industries and make a partnership which infuses such a confidence in them that they will be partners during large scale production.

Our previous experiences show that it will be extremely difficult to give the total production responsibility to one industry, which can coordinate with all other agencies, integrate the system and deliver to the user. Hence, it was decided to set up our own integration complex and produce all the components and subsystems through industrial network. JV was not provided with funds for setting up facilities at the industries for production of components. Hence, we resorted to an approach of requesting industries to set up special production facility for subsystems like airframe and requested DRDO to set up the integration complex to be operated by the JV. Moreover, being a joint effort between the two countries, considerable amount of coordination was required to put things together in ensuring timely delivery of quality systems to the users.

PSUs don't invest to build production facilities unless special provision of funds is made available to them from the Government. Private industries could invest provided there is an assurance of orders, where they can amortise these expenditure during production. With this approach in mind, the BrahMos team evolved a policy to retain system design, system engineering, system integration and system management with the Company and go for partnership, MOUs with leading industries for production of subsystems, components and materials. The procedure for implementing this policy is outlined below.

System Design

A group of nearly 50 scientists from DRDO and BrahMos jointly positioned at NPOM, Moscow to interact with the Russian Design Bureaus over the design and interface details based on Indian and Russian process stock technologies. A composite preliminary design got evolved which got the approval from the joint team. Inputs from multiple DRDO laboratories helped to complete the design process for missiles. RCI designed high performance, fast reaction Inertial Navigation System, On Board Processors, Missile Interface Units for the missile. The design details on the ground complexes, Fire Control Systems, Launchers, Mobile Command Post and software got evolved by DRDO in collaboration with selected industries. The special expertise available with R&DE(E) for design of vehicles and structures helped to configure the shore complex and ship based installations leading to patenting some of the concepts and designs. In this process, a design centre got evolved at BrahMos Headquarters with expertise in Naval Architecture, Structures, Marine Engineering, Electrical and Electronic Systems, Gas Dynamics, Fluid Mechanics and also in transfer alignments. This capability helped in installation of BRAHMOS systems in number of ships that came for retrofit and to others which were being built under new construction programmes. Another group of scientists undertook the development of simulation packages for war- gaming towards best utilization of the weapon system. Substantial contribution came from the Indian side to the JV resulting in gaining mutual respect between the specialists of India and Russia with knowledge sharing making the JV respectful.

System Engineering

The joint venture became the custodian of the design documents, detailed drawings, process documentation, standards, test methods and reports, inspection and quality procedures, system evaluation and acceptance procedures, flight test plans and test results and all other relevant documents. The JV also became the focal point for joint reviews, coordination with all design and development agencies, working out the interface documents resulting responsible for all system engineering tasks.

System Integration

The responsibility of integrating the subsystems, assembling the final missile, the mobile complex for the land version and ship complex for the naval version has been taken up by the JV. In order to carry out the system integration of the missile, separate facilities have been created with trained manpower and infrastructure at multiple centres. Each technician, supervisor, overseeing engineer and inspector has gone through appropriate training and attestation. All the equipment, tooling, test fixtures and methodologies have also been attested to ensure high quality standards. Each deliverable system has been assured for a minimum of ten years storage life which required meticulous quality control procedures that were evolved jointly by the Indian and Russian scientists with participation of Missile System Quality Assurance Agency (MSQAA) authorized Government Inspectors. By taking control of the entire responsibility, the JV resulted in (a) complete control of schedule of activities, supply chain management, industry coordination, problem solving to ensure deliveries to the end user on time; and (b) quality checks at assembly stage and final checkout.

System Management

The contracts with NPOM, DRDO, Industries and all other agencies were carried out meticulously to detail the tasks, milestones and review methodology to ensure common systems and procedures followed with single point contact. Ultimately, the motto is to meet the schedule of deliveries to the Armed Forces. This was possible only through the integrated management. The responsibility to have system management under JV ensured building confidence with the customers.

BRAHMOS PUBLIC-PRIVATE INDUSTRIAL CONSORTIUM

The concept of implementing role of public-private industry partnership in the production of BrahMos missile is a unique one. Criteria for the identification of Industry partners had been laid out and was carefully reviewed and verified. The criteria for industries selection was based on (a) their willingness to invest own funds to support or sustain the capacity and also to increase quantity on requirement basis, necessitating a modular approach; (b) experience in aerospace technology products; (c) the ability to absorb the state of the art technology with high level of knowledge competence; and (d) high governance indicators. The joint specialists' team from both the countries assessed the manufacturing and production capabilities of various industries and short listed specific industries who could be the partners to BrahMos. This report was presented to the Board of Directors. The Board appointed a specialist committee with Board members to scan through these industries.

Based on the finalization of the list of major industries by the Board, Government approval was obtained for the consortium of public and private industry partnership. The best part was, the industries came forward, enhanced their infrastructure by funding from their own source and manufactured subsystems for an advanced supersonic cruise missile. The industries expanded their infrastructure and made exclusive production set up for the missile subsystems. There are many other small private industries who could either provide support to BrahMos directly or who could become subcontractors to major industries. 20 major industries and more than 200 small and medium scale industries became partners to BrahMos which were evaluated by the Government inspectors and were supervised by the Russian specialists. All these tasks were carried out in record time in a concurrent manner, without knowing the number of units to be produced as there was no production order with us at that point of time. It was amazing to see the overwhelming response of the industries expressing full confidence in the leadership. The industries we so confident that missile would be produced in large numbers and they will not be treated like just other defence production organization during the production stage. In many other cases, the industries participated in development, product acceptance by the user took many many years; by the time, the technology became obsolete; and only PSUs were given the production orders. The industries lost interest in participation in the development of defence equipment. This trend was reversed by BrahMos. Thus, the 'BrahMos Missile Industry Consortium', the first of its kind in India was established. Human Resources of more than

20,000 specialists, technicians and engineers are contributing for the manufacture of BRAHMOS subsystems at different industries. The contractual obligations are fulfilled and industries produce and deliver the products well in time to ensure smooth functioning of the missile integration and its timely delivery to the user. The major industries and the systems/subsystems produced by them are enumerated in the following table. The list is not exhaustive.

Name of Industry	Systems / Subsystems
Larson & Toubro, Mumbai,	Composite Airframe, Canister, Launchers for ships, FCS, Missile components
Godrej & Boyce, Mumbai	Metallic Airframes(Nose Cap, Wings & Fins, CPHS, F3), Launcher and launch tubes/ containers
HAL, Hyderabad & Korwa,	INS
ECIL, Hyderabad,	Communication Eqpt., MCP, MAL integration
Astra Microwaves	Telemetry
Ananth Technologies	OBC, MIU, Telemetry
Data Patterns, Chennai	FCS, SCAN Seeker, COE, ATE, Airborne PCD
BHEL	Launcher Systems
BEML, Bangalore	Vehicles for MAL and MCP
HEB, Tiruchirappalli	Batteries
BATL, Tiruvananthapuram	All metallic parts of the missile including airframe and booster. Containers, Airborne Launcher, etc.

Russian Industries

Contributions from Russian industries came in a big way during the initial stages of the development and production, as our partner NPOM had already created a network of industry partners. This came as a boon to receive high quality products and all of them designed with materials of Russian origin and by following GOST standards. The situation became very critical in the early 1990s due to the formation of CIS countries. Ex owners of talent pool to other countries became a problem for Russia. The cost of production kept

Wing
Fins
Air Frame
GODREJ
C.P.H.S
Nose Cap
ENGINE PRODUCTION
MOBILE AUTONOMOUS LAUNCHER
VEM & BATL
HEB
INS Electronics
HAL & ANANTH TECH.
INS
OBC, MIU & telemetry
RAP
LAUNCHERS
Air Frame Sections
L&T
Radome
Internal Nose cap
FCS
CANISTER
Transmitter
Astra Microwave
Data Patterns

soaring high. Hence, it became necessary to judiciously plan the production in Russia and in India, keeping the equilibrium of win-win situation. The JV approach significantly contributed to the growth of Russian industries by retaining the talent which helped the Indian industries to a large extent for undertaking the production of state-of-the-art technology subsystems.

It was amazing to see that the Russian experts were trying to understand the corporate culture in India. As a result of which, the partners of Russian industries (who were associated with NPOM) became independent corporation(s). The Russian Govt. also gave a rare decision through a presidential decree for the BrahMos JV to directly interact with NPOM and subsequently with other industries to fast track the proceedings.

PO Strela, Orenburg	Ramjet engine, airframes, integration
Granit, St. Petersburg	Seeker, COE
Iskra, Perm	Booster and other solid propellants
Avangard, Safonova	Canister
Mashinostroetel, Perm	Booster
Vimpel, Moscow	Ground Complex

INTEGRATION FACILITY

Along with establishing of this consortium, we simultaneously started establishing the system integration facility at Hyderabad. Today, the integration facility at Hyderabad has become a world class state-of-the-art complex. All these happened in quick succession and we started integrating missiles within India.

Soon, we built another integration complex at Nagpur. This facility was realized to augment the production rate as we have received large orders from the three services. Network of technical positions have been established for storage, maintenance and operations of the delivered systems. A pro-active step has been taken to establish a unique product support centre close to these facilities for real time support and to keep the equipments in ready to use conditions. Today, we have state-of-the-art production, integration and storage facilities.

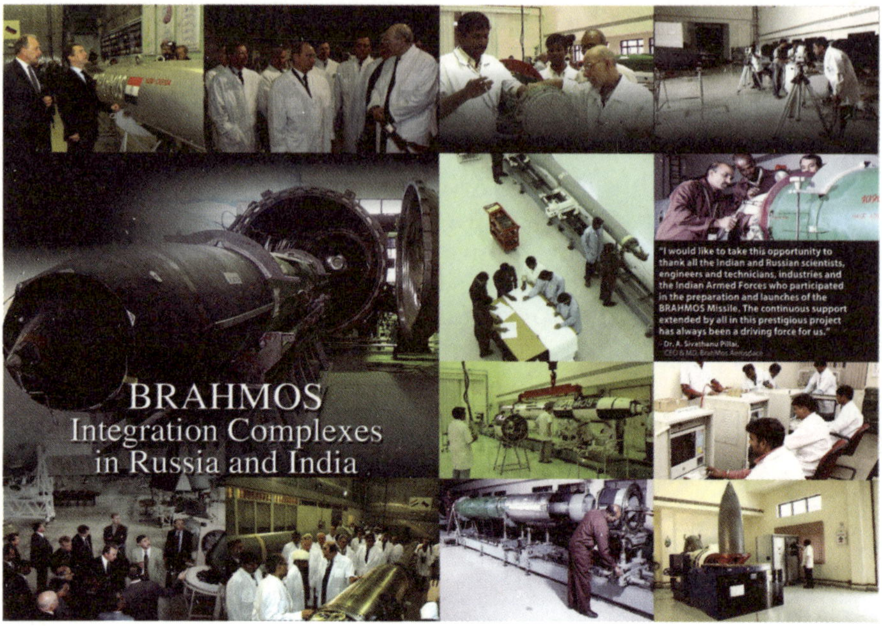

During the initial phase of this partnership, the industries invested their own funds to build up the infrastructure, equipment leading to an exclusive setup for BrahMos. Only limited quantity was ordered for development trials. But during this phase, we could establish the technological competence of the industries. This was done through appropriate training, visits to Russian industries, involving Russian specialists to contribute in detailed drawings, working for tolerances, infusing new technology areas, developing equivalent materials, attestation of the quality level of technicians to undertake special product realization, involving Government inspectors at regular intervals and a large scale qualification tests to prove the subsystems, and so on. Several reviews and interactions have tremendously helped these industries to upgrade themselves, making them committed towards product excellence and the Joint Venture. The development flight trials of BRAHMOS went exceedingly well, giving 100 per cent success to the Navy which resulted in placement of production orders by the Indian Navy in 2005. Further, trials of the land version missile led to the placement of orders by Indian Army in 2006. This brought jubilation to the Indian and Russian industries.

PART 7

In the Horizon – The Future

"When an idea exclusively occupies the mind, it is transformed into an actual physical or mental state."

- Swami Vivekananda

BrahMos Aerospace is a growing enterprise and a role model of high repute, heralding the successful cooperation between the two great nations. The progressive thinking continues to make the organizations vibrant almost every day. In order to keep the vibrancy going for many years to come we have evolved "BrahMos Vision 2050". In this part, I'll be narrating how the future of BrahMos would shape up.

17

VISION 2050

BrahMos Vision 2050

The Bible Proverb 29:18 quotes: "Where there is no Vision, the people perish".. The same holds true for the survival of any organization in the long run. The organizational growth depends on the foundation it has created for sustained growth and vision dovetailing with the growing demand of the customer, maintaining the technology edge with many upgrades. The long term existence of the organisation depends on realisation of newer high performance products and the young energetic team. The superiority of any products over its competitors in terms of performance, reliability, cost effectiveness, production in quantities with quality, longer shelf-life and on time product support fuels in sustenance of organisation's long term viability. New products certainly bring new dimensions to the organizations, thus making it formidable.

BrahMos formed a combined team comprising young energetic engineers and experienced scholars in various fields and evolved a vision document "BrahMos Vision 2050". The vision document has been divided into two portions. First is the BrahMos with current systems and its newer roles while the other is BRAHMOS-II. For the current systems, series of activities such as augmentation of the facilities for production and deliveries to meet the schedule, construction of new product support facility, plan for achieving operational excellence have been brought in the document. Also, it has been suggested to have a new role for the existing system, such as, Air Launched version and the technological upgrades & Modernization of subsystems for Performance Improvements and for undetectability. The second is the realization of Hypersonic version to maintain being the world leader in cruise missile. The document also suggests various upgrades with newer versions, such as, miniaturized BRAHMOS for multiple platforms- aircraft, submarine and ship. One of the other important features of the Vision Document is the realization of hypersonic version - BRAHMOS-II.

There exist a variety of ballistic and cruise missiles with different ranges in most of the countries either developed by them or procured from outside. As described earlier, different propulsion systems are used for different speed regimes (subsonic, supersonic and hypersonic) and different platform systems (land, mobile static, silo, ship, submarine) and missiles are operated in the regimes. India and Russia have together succeeded in developing and inducting the supersonic BRAHMOS missile through the joint venture effort. BRAHMOS has given a big lead to India in achieving the best cruise missile of the world. To keep up the lead in the Cruise missile technology, India and Russia together have agreed to progress on the design and development of Hypersonic cruise missile BRAHMOS-II. Hypersonic technology will be the future in terms of astonishingly higher speed with reduced cost.

Overview of Hypersonic Development in the world

Hypersonics refer to the speed greater than Mach 5, i.e., about 6,000 km per hour. Theoretically, it is possible to reach Mach 25, i.e., 30,000 km per hour in the hypersonic regime. As speed will be the deciding factor for future space travel and reduction in the cost incurred, attempts are being made worldwide to establish technologies required for realizing a vehicle that can fly at hypersonic speed of more than 5 Mach and return as a reusable system. In our book "Thoughts for Change – we can do it" authored by Dr. Abdul Kalam and myself, we have identified Hypersonic as the futuristic core technology for aerospace programme and one of the ten unique technologies which will rule the world.

Hypersonic Research Centre at IISc

To pool the best resources in India and to carry out advanced research in Hypersonics, a centre of excellence has been established at Indian Institute of Science (IISc) on November 2011. The centre has organized an International Conference on hypersonic with participation of several countries to debate over the technological challenges required to achieve a reliable hypersonic vehicle.

France is pursuing its French-LEA hypersonic programme; Japan has undertaken Hytex programme; USA has got hypersonic vehicles like HyFly, X-43, X-51. Conceived in 2004, X-51 has made its first "captive carry" flight in December 2009. The next two flight trials of X-51A

Inauguration of Centre of Excellence in High Speed Aerodynamics at IISc, Bengaluru on 8 November, 2011

ended prematurely. First flight failed due to the combustion gases entering inside the electronic bay and the second flight failed due to interaction of air-intake & combustor flow during scramjet ignition, Third trial lasted only for 15 seconds and lost control due to faulty control fin. The next flight was conducted on May 1, 2013 in which the cruiser travelled about 425 km (230 nautical miles) in just over six minutes reaching a peak speed of Mach 5.1.

In India, DRDO is pushing ahead the development of Hypersonic Technology Demonstrator Vehicle (HSTDV) with an aim to demonstrate autonomous flight of a scramjet integrated vehicle using aviation turbine fuel. Initially, the hypersonic vehicle would fly at an altitude of 30-35 km at 6.2 to 6.5 M with flight duration of 20 seconds. To realize this, technologies being developed at DRDL are aero-propulsion integrated configuration, hypersonic air-intake, hot structures, single expansion ramp nozzle, scramjet combustors, etc. The technologies to be developed for a long duration flight (~600 sec) are Nose tip, wing and tail leading edges, Nickel (Ni) super alloy / Titanium (Ti) alloy honey comb based thermal protection system, protective coating, active cooling of scramjet engine, endothermic fuel, etc.

International Symposium on Hypersonic Aerothermodynamics - Recent Advances (ISHARA)

ISHARA was conducted in Bangalore from 9-13 December 2013, in which BrahMos was one of the major sponsors. The mission of the conference was to understand the complex yet fascinating flow intricacies associated with aerothermodynamic phenomenon in hypersonic flight corridor. Chemically reacting boundary layers, strong entropy gradients in the flow field, massive viscous dissipation, laminar to turbulent flow transition and complicated shock interactions in completely and partially dissociated flow environments are some of the unusual flow physics associated with hypersonic flows. In this backdrop, the international conference brought together experts in the area of hypersonic aerothermodynamics under a common forum. People from India, Russia, Australia, Japan, UK, Belarus, Spain South Korea, Germany and USA presented papers in the conference.

Aerothermodynamic design issues, supersonic combustion, guidance and control, experimental techniques and computational studies with applications in hypersonic flow were the thrust areas of the conference. Understanding the global status of hypersonic research in major academic and research laboratories, exploring possibilities of international collaborations on basic scientific issues associated with hypersonic aerothermodynamics, and identifying important topics for future research were some of the major goals of the meeting. Valuable data regarding force measurements, ignition delay, OH concentration, heat transfer rates, laminar-turbulent transition in hypersonics were presented. The applicability of novel techniques such as force balance, Particle Image Velocimetry (PIV), Planar Laser Induced Fluorescence (PLIF), schlier in high speed flows were discussed. The conference gave an insight of the development around the globe in hypersonics and served as a forum for interaction of the hypersonic experts around the globe.

BRAHMOS-II

Efforts are going on in the area of hypersonic technology. With a speed around 5-7 Mach and operated by a scramjet engine, the BRAHMOS-II will get configured. The challenge will be realization of high temperature materials that would be used in the engine for sustaining long duration operation and in the leading edges of the vehicle and the control structure interaction. Development of scramjet engine is a formidable technological challenge before us.

Scramjet Engine

The engine of the present BRAHMOS-I missile is ramjet where the air rams into the combustion chamber at the speed and gets compressed by the speed itself and mixes with the atomised spray of the fuel. When ignited, the air expands and rushes out of the nozzle thus propelling the missile at supersonic speeds. After the air enters the engine, its speed is reduced to subsonic regime for combustion. In a scramjet engine, the combustion of the fuel and air takes place when the air itself is in supersonic speed. This provides enormous speed to the carrier vehicle. The scramjet engine though looks so simple, is very hard to realize, particularly due to the high temperature combustion process.

A Hypersonic flight within the atmosphere generates terrible drag because of which the airframe of missile gets immensely heated due to the drag induced by air friction. These temperatures can be many times greater than air surrounding it. Again, maintaining combustion in the supersonic flow of incoming air itself is an additional challenge, as the fuel injection, mixing, and ignition cycle have to be completed within milliseconds after the air has enters the engine.

To keep the combustion rate of fuel constant and the missile to have a constant speed, it is necessary that the pressure and temperature in the engine must also be constant else it would not be possible for the missile to manoeuvre. Such restrictions may lead the missile to attain specific rates for climb and cruise.

In DRDL, the development of a SCRAM jet engine is in progress and the ground tests have given good results. Scramjet combustor has been designed with Nimonic C263 material with a maximum heat flux of 150 W/cm^2 and 1400K casing temperature. The ground tests have been conducted for 20 seconds duration. Spontaneous ignition and sustained combustion is achieved under supersonic flow conditions. Performance of the tests has been evaluated using test data and CFD. The efforts have fructified and the engine is capable of reaching a speed corresponding to the speed of the vehicle which is more than mach 6.0. This has been very encouraging. With this, the supersonic combustion has been established. This engine would be used in the hypersonic vehicle and flight tested. This technology would be used as a process stock of India in the JV, for BRAHMOS-II.

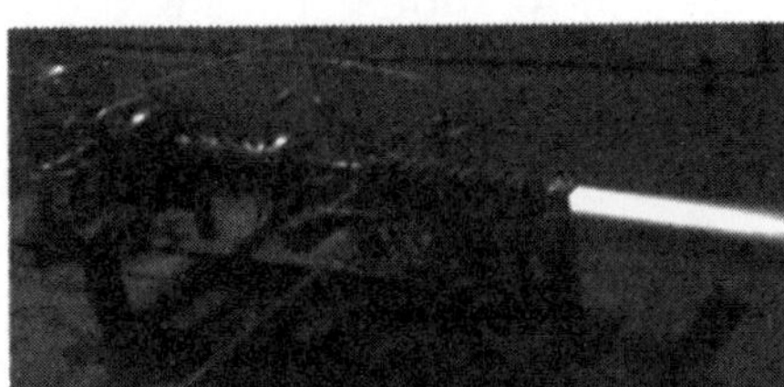

Guidance System

The available guidance systems wouldn't be suitable for hypersonic speeds as the flight vehicle may have to face tremendous vibrations, aerodynamic heating and shock. Though both BRAHMOS-I and BRAHMOS-II have similar missions, but their mission profiles will vary in terms of induced restriction in speed of operation.

Advantages of Hypersonic BRAHMOS

A hypersonic missile definitely has an advantage of speed, as in future wars, more defensive systems against cruise missiles would be realized. Hence, a need to give less reaction time to the enemy to defend himself has arisen. Say for example, an enemy missile base situated at a distance of 290 km can be destroyed roughly in 5 minutes, by BRAHMOS-I. Whereas, the hypersonic BRAHMOS-II with a speed of mach 6 can execute the same mission in around 2.5 minutes thus minimizing the response time by half.

HYPERSONIC REUSABLE CRUISE MISSILE– SUDHARSHAN CHAKRA

After the realization of the BRAHMOS-II, the next task would be to develop reusable version of the hypersonic missile. After this, we would embark on development of hyper planes with multiple applications.

When I think of Hindu Mythology, I think of Lord Vishnu with Sudharshan Chakra portrayed on the index finger of right hand, its technological marvels amazes me a lot. The Sudarshana Chakra is a disk like super weapon with 108 serrated edges which always remains in motion to annihilate the enemy and returns with success back to its original position. Can we think of realizing such a weapon which can fly at a very fast speed, execute the mission and come back and lands on our soil? This terminology is called Hypersonic Reusable Missile.

The present and the upcoming version of the missile fly towards its target and destroy the target along with itself. Such weapons are one time use weapons. That is the purpose of any missile in war. The ideology of Reusable Cruise Missiles is to fly towards its target following a designated trajectory, eject the warhead over the target and then return to its home base. This type of missile would be similar to an unmanned aerial bomber aircraft that would drop the bomb over its target and return to its base. Since the speed of the missile is very high, its lethality is bound to be high. Such a weapon platform will define a new paradigm in warfare without risking the pilot's life and brining its enemy to destruction.

A typical hypersonic air launched missile with reusability flying at an altitude of 30-40 km altitude in cruise mode at Mach 7 is described below. The typical mission would be carried out through an air launched missile at an altitude of 10 km and boosted to 3.5 Mach at 15 km altitude. Kerosene based dual mode ramjet propulsion will take it to 5 Mach at an altitude of 23 km. Scramjet propulsion will propel the missile with a cruising speed of 7 Mach at an altitude of 35 km. The missile would deliver the payload at the designated target and will fly back to its destination.

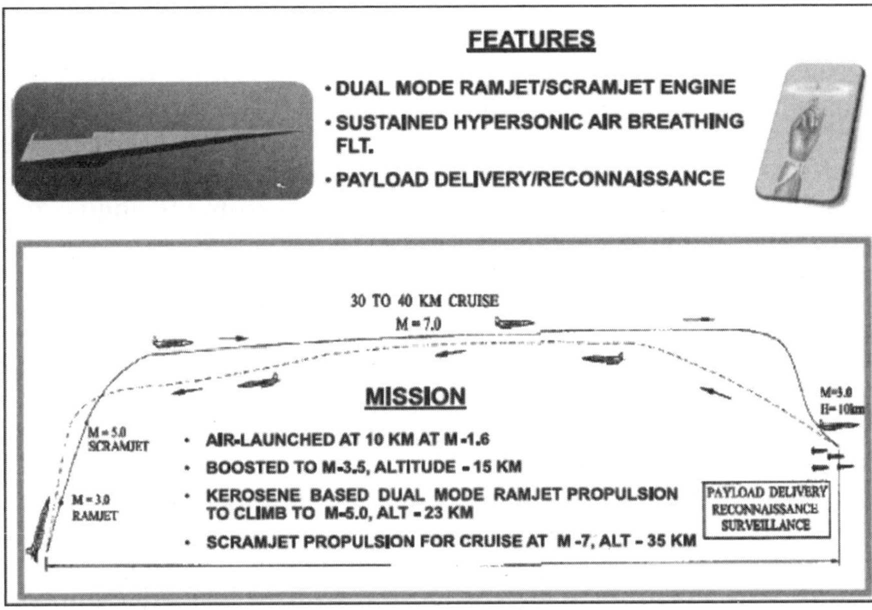

HYPERPLANES

What is a hyperplane? We have heard about Space Shuttle of USA. How can we forget Kalpana Chawla, who along with other crew died at the re-entry stage of the space shuttle? Is that a reusable system? The Space Shuttle can fly many-many times.

You might be aware that the payload fraction of present generation expendable launch vehicles in the world does not exceed 1% or 2% of the launch weight. Thus, to put one or two tonnes in space requires more than one hundred tonnes of launch weight, most of which - nearly 70% - is oxidizer. Such space transportation systems, with marginal payload fractions, are quite uneconomical for carrying out mass transportation and to carry freight and people for the missions of earth, Moon and the Mars. Definitely, there is a need for developing reusable launch vehicles, which can bring down the cost of placing payload in orbit from the present US $ 20,000 per kg to US $ 2000 per kg and eventually to $ 200 per kg.

India proposed in the International Astronautical Federation in 1988 at Bangalore, for the first time, a new concept of Hyperplane. The team led by Air Cmde. (Retd) Gopalaswamy brought out the fact that the payload efficiency can be improved to 15 per cent compared to the 7 to 8 per cent proposed by USA, Europe and Japan. The idea is to develop a hyperplane vehicle that can take off from conventional airfields, collect air from the atmosphere on its way up, liquefy it, separate oxygen and store it on board for subsequent flight beyond the atmosphere. It would take off horizontally like a conventional airplane from a conventional airstrip using turbo-ramjet engines that burn air and hydrogen. Upon reaching the cruising altitude, the vehicle would use scramjet air breathing propulsion to accelerate from Mach 4 to Mach 8. During this cruising phase, an on-board heat exchange mechanism would collect the hot air from the engine and convert it into liquid oxygen. The liquid oxygen thus collected would be used in the final phase of the flight at the time when the rocket engine burns the collected liquid oxygen and the carried hydrogen to reach the orbit. The vehicle would be designed to permit at least a hundred re-entries into the atmosphere. When operational, it is expected to be capable of delivering a payload weighing up to 1,000 kg to low earth orbit. It would be the cheapest way to deliver material to space.

This type of mission will be highly useful for multiple applications. In the case of Hyperplane, the aim is to achieve larger payload fraction. The space shuttle of USA with 2,000 tonnes takeoff weight could launch only 30 tonnes in low earth orbit, giving a payload fraction of 1.5 per cent. India's concept of Hyperplane aims to realize 15 per cent of payload fraction. This increased payload efficiency will considerably reduce the launch cost per mission and will enable multiple missions such as transport, reconnaissance, payload delivery, satellite injection, etc.

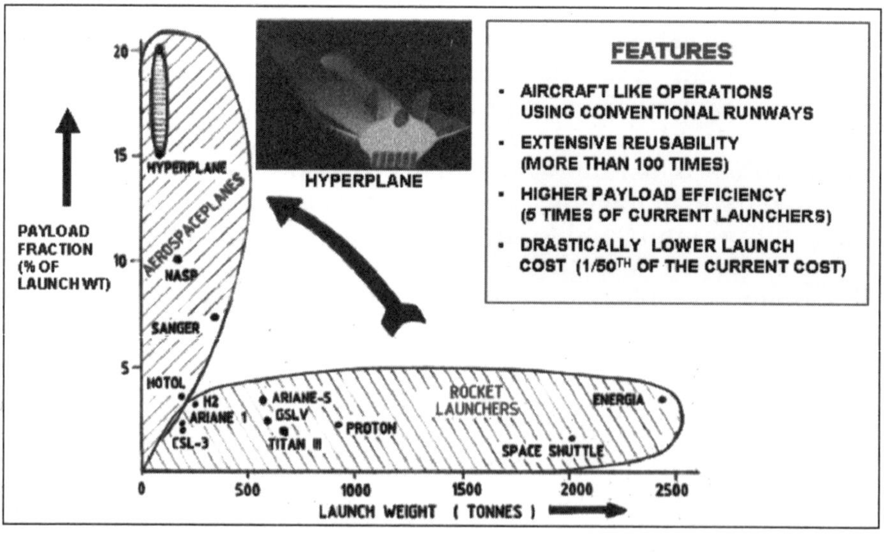

On a typical mission, Hyperplane would take off with 100 tonnes weight using fan ramjet engine and then on scramjet mode for nearly 1000 sec. during which it collects the left over air, cools it and separates as liquid oxygen. This increases its weight to 166 tonnes, thereafter it flies in rocket engine mode using the liquid oxygen and stored liquid hydrogen to deliver a payload of 16 tonnes. This concept of mass addition in flight is unique and has been conceived by Indian scientists.

SPACE BASED SOLAR POWER

The constraints on growth of power capacity like availability of land, water and fuel supply chains (for coal as well as nuclear power) and silting of dams, etc, are among the major factors. These factors have limited the power capacity growth in India to a maximum of about 6 GW/year. While fossil fuels like oil and natural gas are depleting rapidly and available renewable (non-solar) energy sources like wind, biomass, and ocean thermal/wave energy are limited to about 100 GW, India is now concentrating on the terrestrial solar power with its National Solar Mission targets of 20 GW terrestrial solar power by 2020, 100 GW by 2030, and 200 GW by 2052.

However, the availability of terrestrial solar power is constrained to an average 5.6 sun-hours per day in India, i.e., the national solar mission target of 200 GW terrestrial solar power by 2052 is actually equivalent to about 36 GW on a 24x7 basis. Solar energy harvested in space is now recognizable as an option for a 24x7 source of clean, perennial, abundantly available power. The technology for building and orbiting space solar power stations is complex, technologically and administratively challenging, and may take 10-15 years to be comprehensively demonstrated in space and needs international cooperation. It is a minimum technological requirement that the present cost of $20,000 per orbital kilograms has to be brought down to less than $2,000 per orbital kilograms.

There have been several advances in materials and technologies which have greatly enhanced the commercial attractiveness of harvesting energy from space. Considerable advances have been made in enhancing solar cell efficiencies, in reducing their weight and costs. It is expected that by 2020, electricity from nanotechnology-based solar cells in commercial production are likely to be competitive with that from fossil fuels. However, on ground, solar arrays operate only for 6-8 hours in the day and this necessitates the use of SSP systems to meet the base load power demands. In the Photovoltaic Energy Conversion area, the recent research advancement in the CNT based solar photovoltaic cells gives the impressive laboratory results of up to 50% efficiency. Wireless power transmission from space to earth has also shown good results, but the technological selection and its viability on whether microwave or laser power transmission or any other solutions like nano-power pack needs to be worked out and its implications need to studied. Ultimately the overall efficiency of SSP Chain has a determining role to play. Also, it is essential to estimate the space transportation and assembly in space through RLV or ELV.

In India, the need for space based solar power stations was identified way back in 1993 in anticipation of the emerging global energy crisis of the 21st century. Since then, rather prior to 1987, work has been carried out in India on advanced space transportation system design concepts for affordable space solar power. India's vision is to bring down the cost of access to space to less than $ 2000 per orbital kilogram initially.

PART 8

The Lighthouse

> **"I was always looking outside myself for strength and confidence, but it comes from within. It is there all the time."**
>
> **- Anna Freud**

The importance of BrahMos Joint Venture is that it has given gains beyond boundaries placing India in a position of undisputable technology leader. In fact, it is a game changer if we can derive the message correctly and use it for future development and production of defence equipment to achieve our cherished mission of self-reliance.

'Anna Freud', a famous Austrian psychologist and daughter of 'Sigmund Freud', the father of 'psychoanalysis', says "I was always looking outside myself for strength and confidence, but it comes from within. It is there all the time." Let us utilize our inherent strengths to achieve results.

18

WHAT WE LEARN FROM BRAHMOS

Embarking on this joint venture, enormous gains have been achieved by both India and Russia. It is a win-win partnership for both the nations in many ways and it is acclaimed by the top leaders of both nations as the shining example of international cooperation. Not only both the nations have gained in terms of strategic advantage, strengthening the traditional friendship, India in particular has the most potent supersonic cruise missile, the best of its kind in the world, in its armoury. When BrahMos Land mobile version was delivered to the Chief of the Army by the Supreme Commander of Armed Forces, the President of India Dr. Kalam stated, "India has been always fifth or sixth nation in any major milestones we achieved like SLV-3, Agni, Nuclear tests. I was always longing, when India will become the first Nation to possess the most advanced system in the world. BRAHMOS made India proud by making it the first Nation to have a universal operational supersonic cruise missile delivered to the Indian Armed Forces thereby demolishing the Sixth Nation Syndrome." The readers will accept that Dr. Kalam's words of assessment gave a great meaning to the BrahMos joint venture.

Apart from win-win strategy, there were considerable gains for India and Russia through this joint venture. I wish to give the essence of our achievements.

LEAP FROGGING

India was going through a technology transfer phase to produce tanks, aircraft, missiles from other countries thereby learnt little. This is because the product was not the best in the state-of-the-art technology and no design documents were shared through the license transfer. Hence, the learning process was less and all these technology transfers were given to PSUs who have limited technological competence to absorb inputs and utilized them in designing new systems. The indigenous development of systems from DRDO also suffered due to considerable time required for developing newer technologies. This again did not make India the first country to possess the most superior systems in the world. Due to inherent technological gap with developed nations, the efforts we made did not allowed us to break-even with the advanced nations. If we really had to leap-frog, the product should be the best in the world not realized by any other country. BrahMos joint venture proved to the world that here is a product which no one has achieved. Through joint design, development, production and induction in Services, BRAHMOS is a unique product for multiple platforms for multiple missions and for multiple targets. Therefore, India emerged as the first nation to have the best cruise missile in the world.

JV BrahMos is considered as national model for productionization with less investment from the Government side. Sharing of core competence in terms of technology and skill is the basis of formation of JV. Both partners were equally interested in the project right from the very beginning and this interest never faded away. It was a win-win situation for both. India had to increase its defence capability through the induction of most modern system in the fastest possible manner. JV provides around over 20,000 direct/ indirect jobs in manufacturing. In addition, new work culture in design, development and manufacturing including testing, was an outcome of the project. From the Russian side, they needed fund flow for continuation of development programmes. They could get bulk orders for the Russian industries through this missile programme, and enable them to create a corporate culture for managing a consortium of industries. Technologically, they could extend the cruise missile capability for high precision land attack with way points and supersonic steep dive. Keeping this as the motto, both the teams (Russian and Indian) worked with full enthusiasm and conducted the first flight within three years after the commencement of JV. The reliable design and meticulous manufacturing resulted in consecutive successes which further enhanced the mutual trust and confidence between the JV partners and provided them the enthusiasm to continuously work for accelerating the programme. The continuous flow of orders kept this interest further with enhanced dynamism of the team. Regular technical and managerial interactions gave transparency to the JV partners resulting in harmonious relationship between the partners. Vigilant governance of the JV successfully avoided the clash of technological interests by debates and brainstorming sessions and compassion for each other's national and institutional interests. The perseverance and understanding between top management teams helped in solving such conflicts.

WHY NOT BRAHMOS WAY?

We followed and still follow a unique way of management to leapfrog to a next higher level. Through this, we have proved many theories. What are they? We have established a brand that BRAHMOS means Quality.

Novel Concept

The international joint venture BrahMos was formed between two like-minded countries with a novel concept of a JV Model which successfully developed a high end weapon system. This broke the conventional buyer-seller relationship and made both the countries happy to accept that this is a shining example of technological cooperation. This model has been replicated between India and Russia again in the case of 'Medium Transport Aircraft' project and advanced 'Fifth Generation Aircraft'.

Both countries invested almost equal funding of $ 300 millions and the JV is now manufacturing products worth billions of dollars. The time when the Russian forces would start buying BRAHMOS and also when the export orders would be carried out, the returns would further shoot up. If the advantages are crystal clear, why not this JV model is further multiplied to reap more benefits and to gain technological advantages.

Initially, there were many uncertainties and doubts about the joint venture, being the first of its kind. Many impossibilities were turned out to be possible.

1. An international JV with a technologically advanced former superpower.
2. A government owned private company.
3. Public-Private partnership.
4. Fielding a futuristic war winning weapon in a short span of time in India.

GAINS FOR BOTH NATIONS

BrahMos is the first successful joint venture between India and any other country for the design and development of an advanced weapon system. Both countries have invested funds as share capital and technology accrued over 2-3 decades of experience as knowledge capital. The venture is managed by both shareholders through the nominated Board of Directors. The design of the JV product started after assessing the capabilities built in both nations. DRDO has vast resources in terms of experts and technology products. Many successful missile systems were handy to identify the technology products which could be used straightaway.

Definitely both India and Russia have gained through this JV. The gains through this JV are multifaceted. Broadly, the gains achieved can be grouped into Military capability, Technology, Financial and Corporate gains.

Military Gains

Obviously, through this JV, India and Russia got enormous boost to their military capabilities with BRAHMOS. Needless to say again what BRAHMOS is capable of and how it can boost the capability of armed forces.

Strategic Gains

The first important feat is the realization of the most advanced supersonic cruise missile. A missile that was destined to be forgotten at turmoil has been revived and given a new life. The ramjet engine that was developed in Russia was supposed to be left unused due to paucity of funds. By this JV, the money spent by Russia in developing that engine has not gone waste. For India, it could have simply gone outright for the purchase of an anti-ship missile. But that would have simply continued the traditional buyer-seller relationship without taking the time tested friendship to the next higher level.

On the technological front, the missile got a sophisticated INS enabling it to be grouped among the extra high-precision weapons. The simple missile which had to be an 'anti-ship missile has also been converted into a land-attack missile. Even the Russian didn't think of this version and for the first time, world saw a supersonic land attack cruise missile. The usage of Indian INS and software enabled this feat to be achieved. Again, the Onix missile had been conceived as a straight flying missile. But this JV realized the BRAHMOS supersonic cruise missile which can follow waypoints to confuse the enemy and to impart surprise. Notwithstanding these, an air launched version of the missile is also being developed that could be used to hit either enemy's assets on land or ships at sea. So from a scrapped anti-ship missile project we have developed a universal missile that can be launched from land and ship to attack either land assets or ships. A missile that was to serve just one wing of the armed forces is now capable of serving all the three. Upon requirement, both Russia and India can procure any desired version of BRAHMOS for their armed forces. This would result in indirect reinvestment of money on the company enabling it to go forward to hypersonic cruise missile development and beyond.

Technological Gains

On the technological side India gained a lot by understanding the Russian design methodologies, GOST standards and Test & Evaluation procedures. Most importantly without embarking on design and development of an anti-ship cruise missile, we were able to develop multiple versions. The expertise we gained played a crucial role in solving the know-how and know-why issues when augmenting the missile's capability.

Also, India got hands-on over many technologies through this JV. The most notable thing is the usage of canister and nose cap. A missile in the canister with a nose cap is an important feature for the survivability of the missile during underwater launch. Also the missiles can be kept in silos and launched from there. These two launch capabilities form the back bone for deterrence.

How to manufacture such missiles? Their production is not similar to that of ballistic missiles for which expertise is available in abundance in our country. No expertise was available with us for the production of a cruise missile. It is only through this JV project that today India can design and produce a cruise missile because the manufacturing complexities and related remedies have been learnt.

One of the important technological gains is the realization of Quick Reaction INS for BRAHMOS. The reaction time of the Prithvi INS was 30 minutes. The challenges were to reduce the reaction time from 30 minutes to 2 minutes and that the INS to be efficient enough to work in the ambient temperature of -2º to +65º. Several rounds of discussions paved way for the realization of Quick Reaction INS. BRAHMOS being a quick reaction cruise missile, it takes very less time in achieving its mission, Hence, we have achieved the operational readiness by improving the Prithvi INS. BrahMos has succeeded in realizing RLG based INS also. Another important technology that was learnt is the salvo launch capability. This capability has been achieved by the optimization of the INS.

A contemporary Fire Control System has been designed and developed for Pre-Launch Preparation and Launch of the

BRAHMOS missile. The in-built simulators help in providing better training for the users without flogging the operational requirement.

It is through this project that an excellent software code for the seeker and guidance system has emerged which also includes the capability to discriminate its target from a group and precisely hitting them.

We have developed a state-of-the-art mobile launcher. We call it as mobile autonomous launcher because with our ingenuity we made them to operate autonomously. This vehicle is nowadays kindling the minds of many missile scientists internationally to think of. Also, these launchers can operate as a command and control centre in case the command and control vehicle gets destroyed. During war, destroying enemy's command and control facility will inflict a tremendous loss but this would not cause any effect here as any one of the launch vehicle may take the command of the group.

A state-of-the-art mobile command post has also been developed to coordinate and command the mobile autonomous launchers. The command and control systems are entirely developed in India.

Today, BRAHMOS is an offensive missile based on network centric concepts.

Through this JV, a huge "BrahMos Missile Industry complex" has been established. This complex has now attained the capability of manufacturing cruise missiles whereas such an expertise was not available with India in the past. Today, the airframe, INS and avionics, canister, explosives for warhead, solid propellant, MAL, MCP, launchers, ship borne launchers all are manufactured in Indian industries. Apart from these, industries dealing with heavy engineering are now capable of undertaking integration in warships as well. Soon, every component of the missile will be manufactured in India to offer our products at a competitive cost.

Through this JV, Russia has got the Block-II and Block-III supersonic steep diving cruise missile, which was not in its arsenal earlier.

Financial Gains

The BrahMos JV is a wonder in the Indian defence research history. We have been the first in many accords. Huge sum of money is saved both by India and Russia. We got investment capital from Russia which was impossible during those difficult days. Although the required money was arranged from the debt but still it is their money. Most importantly, the money that was supposed to go to Russia was reinvested in the JV. So, without any direct financial commitment at that time, Russia became part of the JV. At the same time, millions of dollars invested by Russia in developing the required technologies has been utilized for the JV Project.

The conditions and restrictions faced by India during weapon purchases have been entirely eliminated. If an anti-ship missile with capabilities similar to BRAHMOS missile were imported, the cost would have been at least two times more when compared to the cost that Indian Armed Forces are paying today and that too without any conditions.

With no equivalent existing anywhere in the world, one BRAHMOS missile completes the job of two to three other subsonic cruise missiles. Any replacement to BRAHMOS would require more units of it to be fired to sink a ship whereas just one BRAHMOS is sufficient to do so. Moreover, the very same missile can engage and destroy land based targets which other missiles are not capable of. So, by paying less money in procuring BRAHMOS from Indian JV Company, Indian armed forces are at an advantageous position of possessing a single missile which is equivalent to the performance of three missiles and which plays role of two different types of missiles, with full product support.

Corporate Gains

Not only these gains, BrahMos Aerospace has made huge inroads as a corporate model - A model that could be followed in the future by other companies. This is because we are the only Integrated Company in the Indian defence Sector. We are the one stop solution for defence needs. How?

TOTAL SOLUTION COMPANY

In India the process of defence research and development is complicated and tedious. DRDO is the design and development agency for weapon systems. DRDO develops the system, tests its efficacy and then proves its capability as required by the user. Then the weapon system is produced by a production agency to which DRDO transfers the manufacturing technology. That production agency may leave the marketing of the weapon system to some other agency. These agencies do not have control on each other. If the user wants a modification to be carried out to enhance the combat potential of the weapon, he has to run between poles to get it done.

A product can succeed only when it is conceived, developed, tested, produced, marketed and supported after sales by a single company. Such a company is called as an integrated company - a total solution provider. An integrated company has inherent flexibilities to make all adjustments for issues that may even arise from an unexpected corner. Be it an issue with respect to production or marketing, an integrated company can handle it very efficiently.

An integrated company for military hardware didn't exist in India ever before and we became the first to do so. In the past, there were some apprehensions regarding an integrated company for defence. Not only that, we have earned respect for India.

In the past, Indian delegates used to throng in international aviation salons and expos, as visitors and potential buyers. But we turned the tables by representing India as a mighty seller competing with many military industrial giants who have been in this sector for almost a century. India was considered militarily weak by first world countries and even the security of the very nation was threatened by an aircraft carrier. But today no naval vessel in this world is capable of standing a chance against BRAHMOS, if they dare to venture the Indian territorial waters with a hostile intent. Because of this superior weapon which no other country possesses, the image about Indian Weaponry has risen immensely in the international arena.

NPOM, which was a design bureau and other companies that produced the missile components have all been integrated into an open Joint stock company. Now NPOM manages the other companies which were previously supplying missile components to it. Because of this JV, all the companies which were on the verge of closing down are now prospering as huge enterprise.

INNOVATION

To achieve things, there should be secret recipes. What are they? Our secrets are simple. A strong long-term practical vision, belief in innovation and following Concurrent Engineering practices. Many innovative practices followed in BrahMos:

- This JV model itself is an innovation, particularly at the right time after Gulf War.
- Going for Supersonic speed regime straightaway is another breakthrough.
- Best of technologies and the expertise harnessed to the fullest advantage to realize a product the most cost-effective manner and at less time.
- While development is in progress, the project was put ahead by going for demonstration launch making the Services to place production order.

- Participation of Indian industries with their investment without any commitment for production
- Conversion of loan repayment money as Russian investment.
- Contracting the developmental works to the share holding entities.
- Universality of the missile for sea and land targets.
- Employment of the missile to attack targets through steep dive.

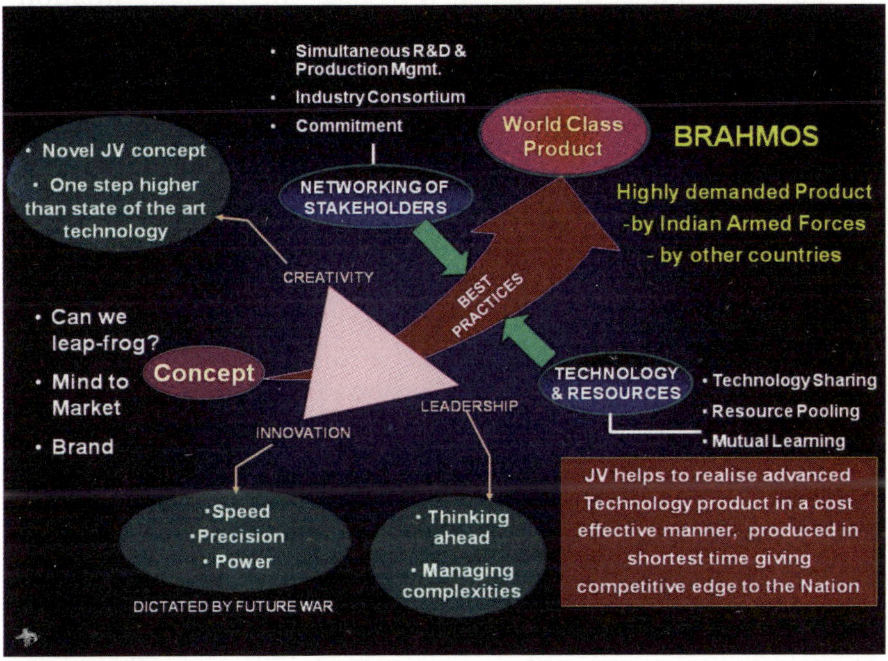

CONCURRENT PRACTICES AND CONTINUOUS UPGRADATIONS

After innovation, the way of our execution is concurrent practices. It is an effective practice through which, the time taken for the product to reach the market from the lab is reduced. I'll try to put it in a nutshell of what we did.

Since the product was seen in the horizon as a successful prototype, actions were initiated to establish the production lines. This resulted in building up of necessary infrastructure for mass production as the prototype was expected to mature as a product after successful flight trials. Then as the orders started to come, we started augmenting the production lines. After a certain number of systems were delivered, we started establishing storage infrastructure for the user. After delivering the product to the user, we went on to establish product support centres. With this approach, we were able to reap the benefits of successfully compressing the time frame.

Taking into account the necessity of continuous product modification and upgrade to always be a market catcher, many versions such as Block-II, Block-III LACM and submarine version have been realized. Air launched version is in its final stages of development. Once we reach this feat, the development of BRAHMOS-II which is taking place at a slower place will be switched to top gear.

Any successful organization must be like flowing water. Water in the clouds has sea as its destination. It falls as rain on hilly jungles. The fallen water collects itself from the droplets to form a small stream. Many streams combine to form a big stream and the process continues till a river is formed. The river flows down to the plains with enormous energy that would eventually make the water to reach the sea. If the river encounters a huge depression on land it first forms itself a lake. It keeps on filling that depression to collect energy. Then, one day finally overflows the depression to continue its journey. On the way it encounters numerous rocky obstacles. It either goes around the obstacle by taking an alternative path or simply blasts away the obstacle with its enormous energy. It merges with various similar small rivers to become a big river and at the zenith it diversifies into branches, nourishes enormous landmass and finally reaches the sea.

RETAINING CONTROL

Another secret is retaining the control through keeping the core responsibilities with us. What are they? Though we outsourced many systems, we kept the System Design, System Engineering, System Integration, and System Management with ourselves. Thus we were able to maintain balance so that everybody gets the benefits.

WIN-WIN SITUATION

A strong bond exists in a JV enterprise where both the sides contributed technologically and financially. JV partners play a complementary role with the spirit of partnership resulting in a "Win-Win Partnership". The gains for the Indian side are (a) a decisive war winning weapon with Indian armoury (which otherwise would not have been available); (b) mobilization of Indian industries for producing the state-of-the-art weapon; (c) ensuring continuous product support to the Armed Forces; (d) best of the technologies available for use in many other projects; (e) saving of huge foreign exchange; (f) becoming an international player for export; (g) giving the status to India as a lead nation among the comity of nations. The gains for the Russian side are (a) support to defence R&D establishments of Russia during the financial crisis due to the disintegration of erstwhile USSR (this helped to control exodus of Russian specialists moving out of the country and also retained their vital technology know-how); (b) Utilization of the equipment for testing; (c) sizeable orders to their industries; (d) continuous interactions and utilization of specialists with newer problems, upgradations and versions; (e) mutual exchange of ideas, review methods, type of equipment with Indian industries providing additional knowledge; (f) newer areas of work - transforming from anti ship role to land attack role, supersonic steep dive, newer versions of trajectories including mountain operations, target acquisition systems, air version, etc., and (g) continuous sustenance of scientific talent for long duration. Above all, both countries gained in retaining the time-tested friendship in an advanced technology programme.

PART 9

Trailblazer

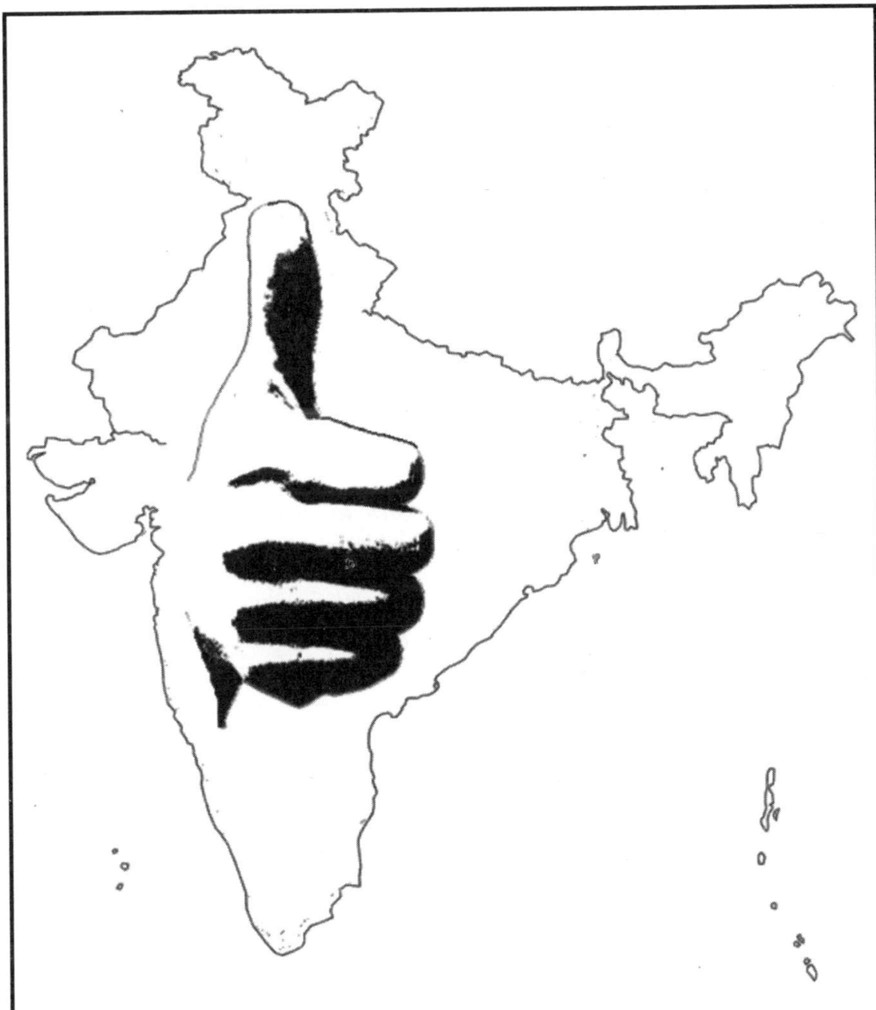

"In modern warfare, a large army is not sufficient; it needs industrial potential behind it.
If the army is the first line of defence, the industry is the second."

- Field Marshal Cariappa

History always gives a chance to learn from past mistakes and wisdom is to be learnt from those mistakes and prevent its reoccurrence. Indians must understand that it is high time for retrospect and to decide the future course which has been knocking at us for quite some time, particularly after the economic revival in 1991 and the nuclear weapon tests in 1998. Why? Because, in geopolitics "Strength Respects Strength - a nation will get respect only for its strength. Only a strong nation can advocate peace in the world." As a nation of rich scientific and civil heritage and being the largest democracy in the world, India could play a lead role as a global leader for the world peace. Strength of the Nation comes from economic prosperity and military power, which in turn comes from Technology and Knowledge to make state-of-the-art products and systems. Military power will only come if we have strong Military Industry Complex which can produce those systems for our Armed Force in achieving self-reliance and to be competitive for export.

19

LEARNING FROM THE HISTORY– SELF RELIANCE

History is the greatest tool of mankind to avoid committing mistakes. The ability of man to learn and apply knowledge has been the key for his survival. Man is a defenceless animal in the biological world. He does not have talons or teeth like predators nor does he possess sharp eyesight like eagles, he cannot see after dark. Man neither has acute hearing nor sense of smell. A defenceless deer can run away smelling the presence of tiger. It is said that animals like dogs, cats and elephants have an extra sense to foresee natural calamities like earthquake and tsunami.

But the ability of man to 'quickly learn' and produce new gadgets made him capable to outwit his predators and dominate them into the forests. From where did man learn? He learnt only from the past and the recorded past is nothing but the history. It is not that only individuals can learn from their past activities and act prudently to avoid the mishaps of the past but an entire nation can learn from its past and act prudently to prevent committing the same mistake again. A nation that learns from the history flourishes and the one does not, deteriorates. For India, it is the 'need of the hour,' to learn from the history, one important lesson which will make India a self-reliant nation.

India was once a land of prosperity and way ahead of other civilizations in the entire world. Great emperors like Asoka the Great, Chandra Gupta Maurya lived and served their subjects with humility. During the reign of Raja Raja Chola-I and his son Rajendra Chola-I, many South East Asian countries and Sri Lanka were under the Chola Empire. Though the Indian Civilization existed as numerous divided kingdoms, growth, peace and harmony existed everywhere. But this prosperous India became a magic fruit of attraction for many invaders of various races starting from Alexander, Timur the Lame

(Tamer Lang), Babur, and Europeans who all looted India's wealth and peace and made Indians as their slaves.

In the hind sight, we can say that the Indian kings were only watching their enemies coming in hordes with sophisticated weaponry. But still they did not researched to get a superior weapon that would deter invasions nor were they intending to fight against the invader. A nation that gave scientific thoughts to the whole world lost the wars only to become slave and slave again. We are yet to come out of that 'mindset'.

On the other hand, the nations that enslaved us have strong industrial base to become economic power. In economic terms, manufacturing is a very important component for prosperity. Therefore, we the Indians must realize that the development of infrastructure and industrial base is vital.

INDUSTRIAL REVOLUTION

Certain instances like the discovery of fire, the discovery of growing plants from their seeds and the invention of wheel were the turning points in the history of man's evolution. Some millennia passed after the invention of wheel without much interesting events. But then came the next revolution - the 'industrial revolution'. When the industrial revolution kick started in the Europe, many new industries were commissioned. Individuals were encouraged to develop new machines to augment production of trade commodities. This mentality did set the phase for the 'research and development' efforts to be carried out and feats once considered impossible became possible at that time, in every sphere. The roots, for the present day advanced technology developments of the west, were laid in this time period. Virtually, machines started replacing human beings in every sphere of production. Not only the commodities, the military weapon systems were also produced through machines. When tried to manufacture arms and ammunitions through machines, people found out that they were able to produce at an unimaginable rate. When wars are fought, the rate at which the fighting forces are supplied with ammunition is a crucial factor for the result of the war. When it was found out that arms and ammunitions could be produced through machines, European Nations, Russia and America established dedicated military industries just to produce arms and ammunitions. Suddenly, even the small countries found out that they were capable of supporting huge fighting forces and became stronger. As a result, their armies and navies started receiving arms and ammunitions at greater rates which made them immensely strong like Israel. Thus, the race to establish, run and maintain military industries began between European Nations.

MILITARY INDUSTRY IN WEST

During the end of the 18th century, military industries became an integral part of the military capability of a nation. These individual military industries, some research capabilities and contributions from the military became a single entity called as military industrial complex which became the second line of defence of a nation. Gen. Eisenhower, supreme commander of allied expeditionary forces during the WW-II and later as US President, first coined this term Military-Industrial Complex (MIC). These military-industrial complexes produced high technology weapons of those eras and nations also continuously upgraded their MICs to attain superiority over the enemy.

During World War-I, it was found by the fighting sides that their war waging capability was actually accounted based on their MIC and not much based on their soldiers. After the WW-I ended; every country in Europe regardless of their political alignment developed their military industry complexes rapidly. Among them, most notably was the Soviet Union. The WW-II broke out and was fought by Germany, France, Italy, UK, Japan, USA and USSR who were the key players of the war. They were the key players only because of their military-industrial capability. Another remarkable nation with enormous industrial capability during this era was the Empire of Japan. To explain the enormity of the potential of MIC we have to look into the WW-II.

During the WW-II, Germany suddenly invaded USSR thus violating a non-aggression pact that existed between them and almost drove USSR to a point of total annihilation. But USSR fought back from that miserable situation and defeated Germany. A crucial move made by the Soviets in the beginning of the war which fetched them victory. Soviets knew that their strength was their industrial might. It was a clear understanding that they could win the war only with their industries. So they embarked on a mammoth transfer of their industries to the Far East away from the reach of the German bombers. Trains carrying the troops to the frontlines were returning with sections of dismantled Soviet industries. It took time for the Soviets to rebuild those dismantled factories deep inside Russia. Until then, Soviets traded time for their huge landmass. USSR started to manufacture the necessary military hardware and supplies. Initially, the production was very slow but it slowly picked momentum and later production went on full steam. They supplied their forces with the necessary weapon systems with a renewed vigour. With this rejuvenation, the Soviet forces were able to push back German forces and reached Berlin, to hoist their 'Hammer and Sickle' flag on Reichstag. Ultimately, the industrial might of Soviet Union became the winning factor.

A look at the unprecedented level of effort for military hardware production of the fighting nations during that war will be interesting. Approximately 280000 Armoured Vehicles (tanks, armoured personnel carriers and other armoured vehicles), 105000 Artillery Systems (howitzers, field guns, rocket launchers, etc), 730000 Military aircraft (all sorts of fighter, bomber, and transport aircraft), 700 warships (including frigates, cruisers, battleships, etc), and 1500 submarines were collectively produced by USA, USSR, UK and Germany.

It may be recalled during the World War – II the first guided missile V-2 were deployed by the Germans. The famous aerospace scientist Wernher von Braun who developed the V-1 and V-2 rockets showed their performance to Hitler. Amazed by the V-2, Hitler ordered its deployment in the war. He asked Von Braun how many he can produce. Von Braun politely told him that he will produce 250 missiles in a year's time. Hitler roared that "you will produce 250 missiles per month from next month onwards. You are empowered to use any industry for that purpose." Germany produced 250 V-2 missiles per month and deployed them. This revolutionized the war theatre with the entry of guided missiles and military industry complex to produce them.

Mankind never saw arms production carried out with this much fury. Just imagine how much steel, coal, diesel, gasoline, and manpower would have been used to produce these many weapon systems. All these steel, petroleum, coal industries need support industries like mining industries, processing and refining industries, etc. Also imagine the machineries required for these industries, logistics, etc. I have not included the trucks, jeeps and other vehicles that also played a huge role in combat effectiveness.

INDIAN SCENARIO

Looking back our own history the rocket production facility was established in the 18th Century by Tipu who deployed thousands of them in the Srirangapatna wars. While we have excellent Government funded Public Sector Units, they alone will not be able to form the Military Industry Complex. Even now the import content for defence system is more than 60 per cent. If we have to increase the self-reliance index, we have to produce weapon and equipment in large quantity in Indian industries. Therefore, the policy should lead to Military Industry Complex for India which should be a consortium of both Public and Private Industries. **India's second Field Marshal Cariappa's words "In modern warfare, a large army is not sufficient, it needs industrial potential behind it. If the army is the first line of defence, the industry is the second."**

BrahMos corporate office is situated in Army Cantonment at Cariappa Marg, always reminds that we are for the customers (Army) and industry support is our strength. We are producing BRAHMOS with 205 partner industries from India and 7 from Russia, thereby creating the 'Industry Consortium" capable to produce weapons and weapon systems in the required level. BrahMos has given a wonderful model of Industry Consortium which can be replicated to bring together both public and private industries to form Military Industry Complex (MIC).

Does any MIC exist in India? The statement may be shocking. But the truth is that we have no structured Industrial Complex. **There is no organised MIC existing in India and there are reasons for that**. The time during which industrial revolution started, India was under the tight grip of slavery. India was perfectly kept away from that industrial revolution, as the British needed raw materials from India for their industries to run. This denial is still a haunt and has resulted in the handicap of Indian industrial infrastructure.

After independence the PSUs were formed for Military Equipment Production. Many private industries have now come to support the indigenous production. The consequence of the non-existence of MIC in its real form is hurting the nation's self-reliance. The MICs of the first world nations are at least many decades old and so their capabilities are a result of these years of experience. Hence, developing and producing high technology military systems is still almost a dream. What about the Indian military requirements?

PROJECTED MILITARY REQUIREMENTS OF INDIA

India is the tenth-largest economy in the world and the third largest in terms of Purchasing Power Parity (PPP) adjusted exchange rates. Due to the large population, on per capita basis, it ranks, 140th in the world or 129th by PPP. Since the economic liberalization of 1991, India's GDP has been growing at a higher rate with a real growth percentage was more than 9% during the year 2009-2010 & 2010-2011. In the year 2011-2012 it was 6.21% and in the year 2012-2013 the real GDP Growth is around 4.9 percent. According to the Planning Commission's report, the contribution of agriculture was 1.79%, industry 3.12% and the Services 6.59%. The decline in manufacturing is alarming and steps are being taken to rectify. The contribution of Services has been growing consistently faster than other sectors. Accordingly to the prediction by Euromonitor International, India would be the third largest economies in the 2020s, measured in Purchasing Power Parity terms. Moreover, India possesses a good amount of best, talented and skilled human resource potential with 65% of the population below 35 years of age. India's defence spending is US$ 41.2 billion in 2012 and is expected to grow at a CAGR of 13.08% to reach US$ 67.87 billion by the year 2016 according to the ICD Research. India's spending on platforms such as aircraft, tank, ship, radar, etc. will be approximately $150 billion in the year 2017 according to McKinsey Analysis.

Not only the military sector but civilian aviation domain is a lucrative market. It is predicted that the Indian civil aviation market will grow rapidly in the forthcoming years thus requiring large number of aircraft to meet the growing demand. Presently airline operators are sourcing aircraft from abroad as there is no indigenous transport aircraft. If an indigenous aircraft is available to the airline operators, then imagine the level of investments saved from sourcing a similar aircraft from abroad. This reduction in investment will also be reflected in the airfare thus providing a highly competitive environment.

In the forthcoming decade to sustain and augment the present growth, India will require enormous energy supplies. It is predicted that India would require 6-7 fold increase in electricity. This requirement could be matched only through nuclear power generation. So nuclear power generation is also a lucrative market for which the Indian MIC can contribute.

How We are Meeting it?

We are almost self-sufficient in strategic military systems - the nuclear tipped ballistic missiles. But strategic military systems are used only for trial and training purposes. They are just deterrence factors, so nobody uses strategic systems. But tactical military systems like tanks, warships, combat aircraft and associated weapons are regularly used to hone the combative skills. So often they need to be replaced by new systems. For tactical military systems, we are dependent only on foreign hands. Today India is the largest weapon importer. But is importing weapon systems good to a nation? Can a nation be ever dependent on other nations? What are the problems?

PROBLEMS IN IMPORTING WEAPON SYSTEMS

Complexities in weapons purchase:

1. The seller will be the master and India will be the dependent.
2. The seller nation's policies affect India: - India has to support the countries that are against Indian policies because of the persuasion from the seller nation.
3. The seller delays and India suffers: - The seller may delay the delivery and India has to pay the price both economically and militarily. After delivering the missiles in time, the seller may delay the delivery of spares and support systems.
4. High life cycle and maintenance cost: - The seller will sell the missile for a cheap price. From the outside, the offer will look like 'Ha! What a missile and that too what a cheap price!' But after buying, then only we will realize that the missile has been sold for a cheap price but the support systems are costly even beyond beliefs. Without the support systems, the missile cannot be maximally exploited. So we have to buy those support systems also!
5. Seller imposes many unfavourable conditions and India cedes: - The seller nation may say that 'the missile should never be fired in war!' Then what for the missiles are?
6. The seller includes delivery charge on the missile's cost.
7. India always depends on the seller for maintenance: - Every time, the OEM has to be contacted and they would send their engineers to our bases for maintenance.
8. The integration works will be carried at the OEM's place in the case of anti-ship versions. For that the vessels have to sail to that place. They have to spend time there for getting integrated with the weapon system.
9. Weapon trials will be conducted at the OEM's convenience.
10. Separate charges might be imposed for testing the weapon system.
11. Then the vessel has to sail back to India from there spending enormous fuel.

We may also face the following difficult conditions due to importing a weapon system.

Inferior Systems

First and foremost is that a selling nation will never sell a weapon platform that is in service in its own armed force. There will always be an export quality weapon system. In non-military business an export quality material is a superior material whereas in military business it is opposite. An export quality weapon system will always be inferior to the ones used by the parent country. If we are buying a military aircraft, critical sensors like radars may be downgraded or associated weapons will be downgraded. No nation sells a weapon system superior or equivalent to theirs. Very rare exceptions could only exist if the selling nation needs funds desperately. But it is very-very rare!

Sky-high Cost

Any weapon system purchased from a foreign nation comes with a sky-high price tag. The system would be inferior to the one possessed by their military. Even that may be acceptable because no nation would sell a system as capable as their's! They follow Machiavelli's ideas. They would have spent considerable money in developing a system. They intend to retrieve that money by selling inferior systems to the importer and that too in the first lot of order. A sky-high price tag for an inferior system and an advanced system for their military at our expense!

Unacceptable Terms and Conditions

Sometimes the seller nation may impose a condition called as end user agreement. That agreement may even demand that those weapon systems, which are meant for offensive operations should not to be used for offensive purposes. The seller country may even send some of its experts to verify that whether we had used the system for any offensive operations!

Unreliable Support During Crisis

In spite of paying huge amount, during the crisis time embargoes may be enforced thus blocking the arterial supply of necessary spares that are required to continue the war effort. India is no new to such embargoes!

Subversive Tracking

There exists a contemporary technology called Radio Frequency Identification (RFID) tags. This tag will emit a radio frequency signal, when it receives a particular signal from a checking system. Each RFID tag can be given a unique code to reply. In such a way, these tags are used to track the movement of goods and serve as an important tool in the logistics industry. But those same RFID tags can be installed in the weapon systems sold by the seller. Then by some clandestine means, those RFIDs can be queried and RFIDs will reply. By identifying the geographical location of the reply, the location of the weapon system could be easily found out. Though any such tracking incidents have not been reported, just because there are no reports, we cannot say that this may not occur!

Sabotage

Assume that we purchase a weapon system under 'End user' condition. Naturally there may be question that, how will the other nation react if that condition is violated. As in the previous point, they can track the location of that system. If the seller senses a violation then, there is a possibility that a small part of the weapon system could be turned off by the seller nation through satellites at times of war. Though such instances have not occurred but there is no way to deny the existence of such possibility also!

Ever Dependent on the Seller

If the purchased weapon system requires a solution for some teething problem that involves a design level modification then we have to depend on the seller. The seller may even put some conditions such that even if with our ingenuity we introduce the required modification we cannot claim Intellectual Property Rights (IPR) for our modifications!

Delay and Poor after Sales Support

If by chance the seller delays the deliveries we only have to pay the price as reduction in terms of fighting capability! There is every possibility that the seller nation says so many stories for the delay.

Indigenous Customisation Impossible

The seller never gives the source code of a critical system's software so that the buyer may not be able to modify the system according to its requirement. Even for this we have to depend on the seller to make the modification, thereby revealing our intentions. Say if we want to modify certain radar to detect some type of weapon platform that the seller has sold to another nation. We have to call the seller to modify it and he will know that we want to detect such and such platform. If at all we get the source codes for a system then that system will definitely not be a contemporary one!

Bending with the Seller Country

Above all, the seller nation's foreign policies will affect us. We may be required to support the countries that are against Indian policies or oppose countries that are friendly to us because of the persuasion from the seller nation.

Job Creation/loss

Recently a new concept called as 'job creation' has emerged. If a selling nation clinches a bulk purchase order, then that nation will generate enormous number of jobs. For example, if we are buying 10 transport aircraft from a country, then some 30000 jobs will be generated in that country. Then imagine the number of jobs that would result from 45000000 crores worth of weapons purchase in the near future. If those purchases are made from India itself then those jobs will be carried out by Indians for Indian companies. Ultimately, the money spent would circulate within India thereby improving our economy.

Corruption, Aggressive Marketing and External Pressure

Corruption is inherent in an arms deal. This exists not only in India but all over the world this is the trend. But in the Indian perspective it has more weightage. When a weapon system is sourced from abroad usually voices raise stating corruptive practices were carried out.

We even have witnessed a foreign company approaching our judiciary for continuing with a particular tender process of our armed force, since the tender was awarded to an Indian Company. Also aggressive marketing by the seller nations to an extent become too much irksome to bear. The seller nations create an extraordinary politico-diplomatic pressure on us to buy their systems.

Myth of Import

Dr. Siddhartha said, "(a) No country parts with the best product. So what is sold to us is only with second grade technology. (b) Design knowledge is not shared and the product is sold under license, recovering the cost spent on development. (c) The seller country controls through spares and critical inputs and makes us ever depending on them".

Do we need to continue import of weapons and equipment forever? How do we improve our self-reliance ratio? How do we mobilize the indigenous talents to produce the best of the products? The remedy is: "Military Industry Complex".

20

MILITARY INDUSTRY COMPLEX

*"It was surprising that whereas every country wanted to produce her own war material, in India, even very senior officials and Ministers wanted to remain dependent on foreign countries and governments for military hardware and would not take any initiative for local production. These people did not understand that a country must not remain forever dependent on another country for her military requirements as, in the event of a war breaking out, that country could stop supplies, putting the receiving country in dire difficulties when her need would be the most acute."**

- Pandit Jawaharlal Nehru

* BN Mullick, "My Years with Nehru 1948-1964", (Allied Publishers, 1972, Page 130)

The first Prime Minister himself realized the need for Military Industry Complex to produce our own equipment for defence. However, still we depend on import for most of our requirement.

There are reasons which are very valid for such a state of affairs. The first and foremost is that developed nations have an industrial set up that is at least a century old. The long time they had been in business has allowed them to create a strong foundation that aids in further augmenting their products' features or introduce a new product in accordance with the wind of trends. But when India became independent, only a handful of industries existed. The number of companies essential for the construction of a modern nation was non-existent. The meagre money available with the government could not make a large industrial infrastructure immediately after independence. Later, the Public Sector Undertakings (PSU) were established by the Government as part of industrial build up.

PUBLIC SECTOR UNDERTAKINGS (PSU)

Virtually for every field, Government established PSUs. For example in the field of public transport sector, state governments formed public transport companies; for air transport two PSUs were started, the Air India and Indian Airlines. For aircraft manufacturing Hindustan Aeronautics Ltd. (HAL) was started. For petroleum exploration and related activities the Oil and Natural Gas Corporation (ONGC) was started. For selling oil and gas products the Hindustan Petroleum Corporation Limited (HPCL) and Indian Oil Corporation Limited (IOCL) were started. For steel manufacturing the Steel Authority of India Limited (SAIL) was started. Like these, today there are around 250 companies existing in India covering almost every commercial sector and under the direction of a ministry dealing with that sector. In the beginning they were essential as there were no other alternatives, as Private sector companies could not venture.

Ordinance Factories and Defence PSUs like HAL, BEL, BEML, BDL, MDL, GRSE, GSL and HSL have monopolies to produce their respective products under license. Capabilities being limited, the import content is still remaining to a larger percentage. Even with the opening up of Indian markets to the foreign companies, the Indian PSUs are not in a position to compete with them. They are surviving just because they are the property of the Indian Government and Indian Government is forced to support them by giving business. All massive production orders are given to them. It is an open secret that the PSUs would never be in a position to deliver the desired output timely and constitute the Military Industry Complex. So what is the remedy?

HOW DID WE DEVELOP DEFENCE INDUSTRY?

Realization of any advanced technology product involves stages like Research, Development, Technology Transfer absorption and production of systems with performance, quality and cost effectiveness. There are three routes- Route A, Route B and Route C to realize technology.

Route A (Know-How) is obtaining the licensed technology with fabrication drawings and the production process supplied by the foreign manufacturer. The concerned industry will realize the product but in this process, industries achieve only limited know-how knowledge acquisition with high life cycle cost. Industrial growth happens only in low technology areas. What we have achieved was creation of large infrastructure, manpower and focussing on the weapon delivery to the user. But we lost the indigenous design capability. Moreover, we have to depend ever on the foreign manufacturers for the spare parts. This approach will not result in acquiring the desired technology from all angles and thus widen technology gap.

Route B (Know-Why) begins with the design and development being done indigenously and the production process done within the country. Industries acquire not only 'Know-How' but also 'Know-Why' knowledge acquisition with low life cycle cost. Here, critical technologies have been realized and many main and allied industries in high technology areas grew. In this process, we had achieved the indigenous design capabilities for combat aircraft, warships and design and development of advanced light helicopters and light combat helicopters having export trust. The main hurdle was the time and cost overrun. Though, there would be initial struggles in this route, the realization of indigenous technologies can take us closer to the world level but would not permit overtaking them.

Route C entails indigenous research and development with limited collaboration in the realization of critical technologies. Indigenous R&D dominates this option. In this process, the development effort & cost is shared between the collaborators enabling fastest realization of the product with the state-of-the-art technology resulting global competitiveness.

Product quality & Value with design & manufacturing

The design plays a vital role in establishing the product quality. If we take the value chain of the product life cycle it can be seen that among the different functions of specification and technical assignments, design, process and manufacturing, production, service and customer feedback and product improvement, wealth generation depends maximum in the design and process & manufacturing phase accounting for 80 per cent of the product life cycle efforts. Design dictates quality, reliability, produceability and cost. Design and manufacturing are powered by knowledge content.

Competitiveness

Technological competitiveness can be achieved through highly competitive products. The dimensions of competitiveness are Product excellence, Cost effectiveness and Delivery in time. The advent of newer technologies and implementation of stringent quality procedures and certification process will definitely lead to excellent products. Continuous improvement of product effectiveness through product upgrades, effective life cycle management and value engineering & cost control measures will result in cost effectiveness. Establishment of production Infrastructure and trained manpower ensures the ability to produce required quantity in time to meet the demand. Application of tools like Enterprise Resource Planning & Supply Chain Management will enhance the effectiveness of the production. Customer delight comes when all this three parameters fall in line.

Examples of Industries Development (Sonar, supercomputer, phase shifter, EW System, etc.)

DRDO has a self-reliance mission with a goal to enhance the production from Indian Industries for defence inventory to nearly 70 percent. With a strong base in indigenous design and development of critical technologies like supercomputer, micro electronics, sensors, high energetics & advanced materials, stealth helped to realize strategic systems. In parallel, industry capability has also been built-up by careful selection of industries for the production of systems by involving them right from the beginning. One good example is building technology capability for sonar. Sonar development encompasses development of various technologies, such as, oceanography, transducers, and electronics. DRDO Laboratory, Naval Physical and Oceanographic Laboratory (NPOL) initially started study of ocean environment and development of transducers and within a decade, started electronics development of small equipment. After achieving enough expertise in these areas, development of full sonar system started in late seventies and first indigenously developed sonar "APSOH" was delivered to the user in early eighties. Sonar Design to production approach by involving the Navy right from the beginning and the industry identified for production was Bharat Electronics (BEL).

The laboratory adopted the system development approach such as "Laboratory Prototype - Design and Engineered (D&E) Model - Production Model" when the sufficient expertise was not achieved. Prototype was made by the laboratory which was subjected to various technical trials. On successful completion of technical trials, the D&E model was produced by the production agency and the model was subjected to user's trials. On successful completion of user's trials, the production model was produced by the production agency. After gaining sufficient amount of expertise and success, D&E model was produced directly, which was subjected to user trials and on successful completion trials production model were supplied by the production agency to Indian Navy. In the later stage, the sonars were produced directly by the production. After attaining the maturity in these technologies, the laboratory developed multiple types of sonar which are being produced by the Indian Industry. Active and continuous involvement of Navy and joint effort of NPOL & BEL from development to production and induction enabled achieving indigenous capability for Sonar Systems.

Another example of design to production is the Samyukta Electronics Warfare system. Defence Electronics Research Laboratory (DLRL) being a prime R&D Laboratory among many other DRDO laboratories. BEL was the prime production agency with the participation of number of Public and Private Industries. Number of academic institutions participated in the development of systems. Strengths of development and production partners, academia and the QA agencies had been synergized. Steering committee and continuous participation by the Army and effective quality monitoring paved way for the successful development of the system.

In the Integrated Guided Missile Development Programme, many critical technologies have been developed and many industries have groomed.

Concurrent engineering practices were used in the development of Prithvi surface to surface missile. There were about 200 technological packages out of these 20 were critical. Various sub-systems were identified for the development of missile and the design and development of these subsystems started at the various work centres simultaneously. The Production Agencies (PAs) were identified for each of the subsystems. One of the PAs was made responsible for the overall integration of the missile. Design of the various subsystems was reviewed at the various levels. Critical design review of these systems was carried with the experts drawn from design agency, production agency and academic institutes. Based on the recommendations of the Critical Design Review Committee, necessary design modifications were carried out by the production agency. The user was involved right from the very beginning of the design inception. Inputs provided by the user were incorporated. Users' inspection agency also participated in clearing up to components level. Various subsystems were assembled at the integration centre. The missile was integrated and subjected to users' trials. On completion of rigorous users' trial, the missile was accepted for induction into the armed forces. With adaptation of concurrent engineering philosophy, we were able to save 5 years in developmental efforts.

When we started development of the medium range surface-to-air missile AKASH, two important technologies dictated the development time. One was the solid propellant based ramjet propulsion to achieve a specific impulse of 600 units twice of a normal solid propellant system. The second one was a command guidance integrated with phased array radar to achieve multi-target tracking and guidance capability. This means that the ground radar needs to lock on to multiple targets and guide multiple missiles towards intercepting those targets using electronic switching of the beams. Design and development of the Radar system was entrusted to a DRDO laboratory. This radar has multiple arrays, each array with several thousands of ferrite phase shifters at different frequencies. The radar beams are steered and controlled electronically in space by programming the Phase Shifters through computer controlled logic. The critical area for development of this Radar was a C-band ferrite based phase shifters for electronic beam steering. Only two countries in the world possess the required technology which was fully guarded. No expertise was available in the country either in design and development or for production of the system.

The most critical thing was the development of ferrite phase shifters. Design and development of the phase shifter was entrusted to Indian Institute of Technology (IIT), Delhi because in the entire country only Professor Bharti Bhatt and Professor Kaul were the two academicians who knew about phase shifters. Now, the requirement came to build a laboratory around the Professors with necessary infrastructure for the design and development of the phase shifters. Though it was a difficult task, it was realized in time. Solid State Physics Laboratory, another DRDO laboratory helped in development of specialty material and CEL a Public Sector Undertaking assisted in the production on receiving technology transfer from IIT. The empowerment of the institution to develop the new technology was accomplished through continuous reviews and help from the R&D labs and concurrent productionization support by an industry. This resulted into the development and establishment of technology in a short period of three years and production commenced from the fourth year.

Graduation of Indian industry

In the early 60s, through technology transfer, industries got process for fabrication, equipment and training for the manufacture of the system. But with DRDO initiated programmes, like sonar, EW and IGMDP, the industries slowly graduated to the level of evolving and manufacturing. During 90s, industries attained the stage of build to design, i.e., the design by the DRDO, Indian industries were able to develop systems, such as, light combat aircraft. In another decade, industries grew to the stage of build and design with the capability to build to specification. Thus, a matured defence industry base with both public and private was formed thus becoming a strategic partner to R&D.

THE BRAHMOS WAY

In the IGMDP era, many industries started building up core competencies in production of ferrous and non-ferrous alloys and composite materials, precision fabrication of components and large sized structures, electronic systems, avionics, sensors, radars, chemicals, cables, connectors, etc., from raw materials stage to systems. The industrial base therefore was strong enough for better utilization in the realization of BRAHMOS missile and ground systems. We needed to build around the capability and suggest augmentation of capacities of these industries and make such partnership which infuses such a confidence in them that they would be partners during large scale production.

Through our past experiences we were aware that it would be extremely difficult to give the total production responsibility to one industry, which can coordinate with all other agencies, integrate the system and deliver to the user. Hence, it was decided to set up our own integration complex and produce all the components and subsystems through industrial network. JV was not provided with funds for setting up facilities at the industries for production components level. Hence, we resorted to an approach of requesting industries to set up special production facility for subsystems like airframe and requesting DRDO to set up the integration complex to be operated by the JV. Moreover, being a joint effort between the two countries, considerable amount of coordination was required to put things together to ensure timely delivery of quality systems to the users.

BrahMos Formula

- Robust Design with unique world class capabilities (Best in the World)
- Systematic validation of systems, Simulation, Checkout procedures and reviews establishing reliability at every stage of development
- Incorporating the lessons learnt through earlier experiences of Prithvi and Agni, and thinking a step better in the product performance
- Continuous performance improvements, upgradations and new

versions for Customer delight, utilizing the experts from India & Russia

- Strong customer loyalty to the Brand
- Consortium of industry partners keeping design, integration and management with JV. All stakeholders networked with integrated management system and transparency in operation
- Faster execution of the tasks due to autonomous operation of JV & empowerment
- Utilization of young energetic talent force with Pride in working for world-class missile for the nation

Shaping the procurement policy by Government towards Self Reliance

Towards establishing the self-reliance in Defence, Government of India has shaped up the procurement policy in its recent Defence Procurement Policy (DPP) 2013. The main objective of the DPP-2013 is to increase the indigenous content by boosting the Indian Defence Industries both public and private. Capital acquisitions are through three types: (i) Buy (ii) Buy and Make (India) and (iii) Make. Buy category means outright purchase of an equipment. It emphasizes for the minimum of 30% indigenous content on cost basis. The offset helped the participation of a large number of medium scale industries. Acquisitions covered under Buy & Make (India) covered systems developed by a foreign industry & technology provided for production by Indian Industry. This must have minimum 50% indigenous content on cost basis. Acquisitions covered under Make (India) would include production of strategic, complex & security sensitive systems (No possibility of import); systems designed, developed and produced indigenously (DRDO + Industry or Consortium of Industries or One Industry); and systems jointly designed, developed & produced under JV (e.g., BrahMos). A minimum of 30% indigenous content on cost basis. Thrust given on the offset policy with increased indigenous content on the production of systems will increase the participation of more number of Indian Industries in the national development.

Present Ambience and Requirements

An important factor prevailing in the present context is the absence of coherence and discrete relationship between the stakeholders viz. Government; Armed Forces; R&D Organizations and the Industries. This resulted in the large import of weapons and equipment due to which there has been continuous dependence on other countries for spares / equipment; there is no effective coordination between R&D organization and Industry as a single entity; and absence of "Mind to Market" thinking.

There are both advantages and disadvantages in all these four sectors of stakeholders. Identifying the bottleneck areas and resolving them and by unifying the strengths of the stakeholders will bring synergy in developing products in the niche areas of technology. Let me narrate the strengths and weaknesses of the stakeholders in the succeeding paragraphs:

Government has got progressively improved Defence Procurement Policy to boost the indigenous production of systems and equipment. Moreover, the offset policy guidelines outlined will induce active participation of industries in the development of systems. More on the benefits of the procurement policy is outlined in the earlier paragraph. What Government can do further is that the thrust should be given on the Long term Defence Strategy and Vision for Defence Industry growth, involving large number of Private Industries as production partners. Government should open up the access for renowned global Companies for having partnership with Indian industries to result in the flow of new manufacturing techniques and state-of-the-art technologies to share with our industries. This will in turn result in the realization of high quality technology products. Allocation of more funding by Government for R&D & infrastructure for establishment and nurturing of advanced research labs and Foundries will strengthen our indigenous R&D capability to help the industries. As in many countries, Legal provision for "Use only Indigenous Weapons" will definitely necessitate the indigenous industry growth and their investment in R&D and go for JVs in high technology products. Defence PSUs, OFs and major identified private industries should become a single entity.

Our highly committed Armed Forces need contemporary, state-of-the-art weapon system to fight against the enemy in the process of protecting the nation. No country gives the best of the weapon and equipment to other countries. When it comes to the development of system indigenously using home grown technologies, there must be level playing acceptance procedures for these systems. At the time of formulation of Qualitative Requirements (QR), user takes the requirements from the existing best systems and adds tough specification to that. The QR once formulated should be the final and there should not be any change in the midway. This will definitely affect the schedule since change in QR affects the entire chain right from the design to process to production to validation. The change in the trained manpower affects the usage of system by the Armed

Forces personnel. In the case of BrahMos, the Armed Forces have raised a separate regiment for the BRAHMOS. Trainings are imparted to the crew at different levels to familiarise with the system and ease of operation. Frequent change of personnel will result in the delay in operation and extended man efforts in providing training to them.

DRDO has got the best brains that puts in all-out efforts to succeed in various missions and development of defence systems. In these five decades, started with the basic research, today, DRDO is capable of building bigger systems and has attained the technological maturity. So far, Rs.170,000 Crs. worth of defence products have been produced and delivered to our users. DRDO has got an excellent track record with success in focussed mission mode projects. As the technologies become obsolescent, fast and proactive adaptations are required to sustain in the rapidly changing technology environment. Technology denial regimes added to the problem of restricting in-flow of technologies. With limited resources and manpower and many projects on hand caused non completion of projects in time. Procedures should not become hindrance to the completion of the projects within time. Judicial allocation of resources to multiple projects will resolve the problem. Giving thrust on Quality of Technology transfer documentation and productionization aspects will accelerate the smooth transfer of technology to the industries.

Large industrial infrastructure is available with Indian Defence PSUs for specific product lines to support Armed Forces (Su-30MKI, T-90S, Radars, Missiles and Ships). At the time of establishment, the main motive was to produce defence systems to cater to the immediate requirement, looking for technology inflow from outside. This has resulted in technology know-how and not the technology know-why due to low absorption of technologies. Low R&D effort in the industries has further worsened the situation. Technological obsolescence and the non-upgradation of machineries suiting to the modern manufacturing techniques led to low productivity and quantity production which has resulted in the high cost for products. Moreover, dire requirement of equipment and systems for Armed Forces forced them to go for import due to the exorbitant delay in delivering the products by the industries. There is tremendous growth with the big business houses by the private players. What is required is the fast absorption of Technologies and increased level of R&D effort. Participation of more Private Players through attractive policies ensures competitiveness.

Military Industry Complex that connects all these stakeholders to take the best out of all is the requirement of today.

Need of the Hour – Military Industry Complex

The need of the Hour is to establish a Military Industry Complex (MIC) at the national level enlisting large and medium industries to be the partners along with Defence PSUs as members of MIC. Procedures are to be formulated which will enable participation of cluster of industries to respond to RFP to design, develop and produce the systems (Irrespective of Private or Public). More thrust has to be given for Government funding for R&D even to Private companies to strengthen the indigenous R&D capability. Regulations and control procedures like USA are to be implemented in managing the private industries for manufacturing of defence systems. Creation of an MIC Authority helps to oversee building up production capability and capacity within India. Encouraging high technology tie-ups / JVs between Indian and other global defence industries will achieve not only competitiveness but also envisage the product for export. India cannot afford to lose any more time in pondering the issue. Indian Military Industry Complex with MIC Authority as regulatory body is an absolute necessity.

This will change the perception of the world towards India from an Importer Country to an Exporting Giant.

PART 10

Reaching the Destination

"Always remember that your present situation is not your final destination, The best is yet to come"

- Anonymous

Right from the beginning of any voyage till its destination, many hurdles – known and unknown can come on the way, but it is the captain who commands the ship to make it a safe journey. Likewise, the creation of any organization and its emergence to success with flying colours is the responsibility of the leader and the great team, nurtured by him. The voyage continues to further missions.....

21

A WORD FROM THE CAPTAIN

During one of my visits to NPO Mashinostroyenia, our partner organization in Russia, I was surprised to see a picture in the room of Dy. Director Generals and other key managers. I was shocked to see my portrait as a captain standing on the ship called BrahMos with the hard work of our Russian friends. Whereas, I have always considered myself as a team member and a bridge between the Indian and Russian Scientists in getting things done based on mutual understandings and discussions at every stage.

Got with the surprise to see the same picture in many rooms, I asked them "Why did you make a painting of this sort and displayed in most of the rooms?" They replied that "Everyone in NPOM is grateful to you and this picture reflects that spirit, we are working for a great mission and we respect you". It is a fact that BrahMos is own lifeline and has a leader like you who has been the driving force to demand and deliver. This is the key to success and we cherish this relationship. In their word, "Dr. Pillai, you are highly respected in NPOM and in Russia for this venture. We will fully support with hard work to ensure that the JV realizes all its goals. This picture is the reflection of that fact." I became a small person in front of their broad mindedness. My respect to Russian friends multiplied. I realized that utmost honesty of purpose and goal clarity have motivated our teams in India and Russia to give best results to the extreme satisfaction of customers, involving them as partners to success. How can we motivate the Indian teams to perform the way the Russians have expressed?

Talent pool

The most important aspect is the management of talent and continuous motivation to perform. The best talent pool is the toppers of the Universities, who aspire to go abroad for higher studies, outside India. If we can encourage the talent pool to work for the organization, we can achieve the most important requirement of youth power to remain long to carry forward the task. This was done exceedingly well in BrahMos to keep the best of manpower as the talent pool with an average age less than 30. In BrahMos, every year, 20 to 25 rank holder students from the Institutions of National Repute are absorbed through campus recruitment. They are given six months training which includes orientation programme, visits to Laboratories and Industries to acquaint with the recent developments of the Project. They are given an independent responsibility of developing a futuristic system. The place becomes ideal for them to utilize their innovation and creativity. 10 years down the line, this precious human resource becomes the part of management and takes the project to newer heights. This is a remarkable success mantra for BrahMos.

Creating Centres of Excellence

Researches in the academic institutions have always been a support for design and development. Academic institutions are the rich source of knowledge and it is necessary to associate the research component with R&D. I have already explained the invaluable contributions of academic institutions. One such example was the successful development of Phased Array Radar and its indigenous production in India. Many institutions had been involved in development of critical technologies. Keeping this in mind, the excellent contributions of the academia, we have created Centres of Excellence to undertake studies related to the futuristic technologies involved in development of advanced weapons. The strength of the Academic institution is identified and based on that Centre of Excellence is established to develop critical technologies that are needed for the realization of technologically advanced products. They become the partner in the development.

Partnership Forums

Be it individuals or industries, recognition is vital for the growth and to keep up the spirit. Forums are the platforms for the partners to express and share their views and allow them to take the ownership and increase the confidence they have in the project. Particularly, in the development of high technology products, associated industries who have invested their money, time and infrastructure naturally expect to get orders placed for them. Once the master organization has kept their words and placed the orders on the industries that were identified for the development of systems / subsystems, their enthusiasm and the confidence increases.

BrahMos on 19th February 2013 organized a meet named as "Ardhik Diwas" (Partnership Day) celebrating 15 glorious years of friendship, trust and achievements marking the signing of Inter Governmental Agreement for BrahMos joint venture. On this occasion, all who have contributed in the formation of joint venture and the personalities who have contributed for its growth and the users who have facilitated the Project were felicitated. The contributors shared their good feelings taking ownership of the project. Hon'ble Defence Minister during his Chief Guest Address said, "This unique BRAHMOS missile in real sense a Brahmastra". During the occasion, a panel discussion was also organized. The panellists include bureaucrats, technocrats, Service personnel (Users) and from media. The panel concluded that BrahMos JV is globally accepted as a Role Model & is recommended for future JVs.

On 30th July 2013, an Industry Meet was organized at Hyderabad to highlight thc vital contributions made by public and private industries in making the BRAHMOS missile a world-class and formidable weapon system and to recognize the contributions of 205 Indian Industries and 7 Russian Industries who are associated with the Project. This generated a positive feeling among the industry partners and made them proudly realize that they have contributed in producing the world class weapon. BrahMos felicitated the Industry partners, laboratories and key individual contributors involved in the design, development and production of BRAHMOS weapon system. The Industries who have participated in the Industry Meet had expressed that by being associated with BrahMos, the status of their technology maturity has increased. The amount of dedication showed by the industries towards delivering high quality systems/subsystems is commendable. The grand event was graced by the Former President of India Dr. APJ Abdul Kalam who termed the BrahMos Joint Venture as a fine example of role model of courage and excellent leadership.

Such forums boost the morale of the participating industries as well as the people who are associated with the project to take the ownership and act as a catalyst in the contribution to perform more.

Steering Committees with the Customer

Steering Committees were one of the influencing factors for the success of the project. Separate Steering committees were constituted for Indian Navy and Indian Army. These Steering Committees are chaired by the respective Vice Chiefs, members from Armed Forces and BrahMos. Periodical reviews were undertaken to bridge the gap between the developer and the user. The decisions required from the User side were raised and clearances obtained at the highest level. This too facilitated the developer, R&D organizations and haled them to be staged on common platform to understand the system requirement. Some of the activities like installation of launcher and associated systems require extensive participation of Indian Navy and there were many hurdles which tried to affect the progress of work. Interventions and decision making at the highest level helped the project to accelerate and execute faster and induction of the system in the Armed Forces.

Enterprise Resource Planning

For greater operational efficiency, there is a need for rationalizing the business processes. Automation and seamless integration of the current processes is an absolute requirement and necessity to obtain efficiency. ERP –Enterprise Resource Planning is a software tool that helps an Organization to manage the business processes which include Product Planning and Development, Manufacturing, Inventory management, etc. It is an integrated system that operates in near real time with a common database supporting all applications. The advantages of ERP are many. It greatly improves the quality and efficiency through smooth running of internal business process resulting in more productivity thus leading to better customer satisfaction. It enhances the organization business by allowing the top management to make critical decisions. We at BrahMos have implemented an ERP solution that brought a big change, majorly in areas like production and product support. BrahMos always stands for the timely delivery of the system to its users thereby maintaining customer delight of high order.

Key Strategic Issues in managing JV

Following are the key strategic issues in managing this Role model Joint Venture:

- Creating and sharing enough value for the partners, maintaining the win-win strategy
- Maintaining enough strategic consistency and compatibility between the partners
- Containing the complexity of managing vital issues with the joint leadership

- Establishing sequence of partnership related to credibility, criticality, speed of growth, marketing the product.

For successful JV and partnership, the following four ingredients are necessary: (a) the ability to maintain faith in each other; (b) the ability to express disagreements with each other in a constructive way; (c) the ability to put the teams together; (d) the ability to complacent each other to account if they are appearing short.

Corner stone's of Success

"Try not to become a person of success but rather
Try to become a person of value"

-Albert Einstein

The result speaks for (a) absolute focus on visible results; (b) increase the pace of tempo and motivate the team to perform; (c) engage in developmental activities for continuous improvement of the product to keep customers' interest; (d) take complete responsibility for the risks in decisions; and (e) show in every step absolute honesty of purpose with ethical standards and values.

Every milestone we crossed was an "Agni Pariksha" and we as a team converted the problems to showering rose petals. In motivating our teams and making our customers happy. The management practices demand '3Cs – Communication, Commitment and Coordination'. Results take centre stage in mobilizing the motivated team than hierarchical command and control. After all, it is the delivery that counts in management and leadership. The most important points to note are (a) visible focus on results; (b) work plans and responsibilities to achieve; (c) constant review and problem solving; (d) tuning the organization structure to meet the defined tasks; (e) networking with partners to outsource more; (f) total control on system design, system engineering, system integration and system management; (g) encouraging employee developmental activities with good HR management; (h) instituting appropriate performance measurement system and above all (i) expression of courage in decision making.

The organizational capability needs planting seeds for a sustainable competitive advantage, means creating a body of knowledge, organizational intelligence, domain expertise and experience in new products, processes and technologies and bringing successfully a new product to market. Constant product upgrade and the birth of new versions with increased capabilities help in maintaining the competitive edge and superiority in the market. Our persistence has always motivated to keep going even after setbacks and has immensely contributed in converting infinite road blocks into opportunities. Novel concept of leapfrogging, unique visionary thinking and the strategic leadership with a determination of getting things done are the success of the joint venture.

Dr. Abdul Kalam always encourages the students: Have a great aim, work hard, acquire knowledge and inculcate such perseverance which empowers you to overcome the challenges. This advice suits any organization as well. Hence, a great vision where employees believe and get themselves committed to achieve, with hard work, build a knowledge organization which excels and have leadership with courage to defeat failures.

When I was in Harvard Business School, Sam Walton gave a lecture. He said: "Commit to your business, believe in it more than anybody else, put deep thoughts in it, live in it and overcome the shortcomings by sheer passion". Because of risks in decisions, entrepreneurship is not for the faint of heart. Cowards do not become good leaders. One has to be bold enough to take risks.

CONCLUSION

People relying on news programmes and newspapers for updating their IQ may not be aware of such experiences. They may never know the amount of pain, sacrifice, dejection and tragedy that have been faced and suffered by mission teams working in our research establishments behind their successes and failures. People generally comment "why these research organizations are not delivering anything rather they are eating our money, we pay as tax?" but I want to reiterate that our scientists and defence personnel are no inferior to anybody. They are putting their best efforts. The purpose of the book is to highlight such sacrifices and rejoices that may always lay hidden from the limelight. We were plundered by barbaric invasions; our wealth was eroded and kept as exhibits in museums; our traditional education systems were uprooted; we were purposely kept away from the industrial revolution; our country was meticulously divided by the colonial rule before it left. But we are regrouping, a new generation is ready and I see a strong and self reliant India in the horizon.

இடும்பைக் கிடும்பைப் படுப்பர் இடும்பைக்

கிடும்பை படாஅ தவர்

Successful leaders can never be defeated by problems.
They become master of the situation and defeat the problems

WORDS OF WISDOM

India is a great nation - A powerhouse of knowledge. We all belong to a great gene of civilisational heritage. India's requirements are large and are steadily growing. To meet out these requirements, to become a global leader and to be among the top three countries, we need organizations and industries that are of world class. We must take on and grow further from this strongly created foundation. Young scientists and technological community must take on to make India stronger and stronger. Nothing is impossible.

So every youth of India must keep this in mind,
"If we are expected to Achieve Results never before Accomplished,
We must employ Methods never before Attempted,
But with utmost honesty"

A. Sivathanu Pillai

Epilogue

Product Competitiveness is a very important aspect in the globalized market. The competitiveness depends on three important factors; (i) the product excellence, (ii) the cost effectiveness and (iii) availability in required quantity and in product delivery in time. If we have to ensure the customer's delight, the three aspects are essential.

Product Excellence means Performance, Quality, Reliability, Newer Technologies, Quality Procedures & Certification. In Cost Effectiveness when we consider the cost we look at continuous improvement of product effectiveness, Effective Life Cycle Management, Value Engineering & Cost Control measures. Availability in Time is very important. After making the system ready if it is not available to the requirement of the user, then whatever effort we have put in is not going to be fruitful. Availability in Time requires production infrastructure, trained manpower, Contractors Management, Enterprise Resource Planning & Supply Chain Management. All have to be planned during the initial stage of the programme. Ultimately DELIVERY COUNTS.

The various options of realizing the product are:

Route A is through the Technology Transfer: For example the Mig-21, we have gone through the license production but with limited know-how & knowledge acquisition. It has resulted in repeating the production of system in India using kit of parts and we have to still depend on spares from other countries.

Route B is totally indigenous. This also we attempted through Prithvi and Agni. Here, we are aware of the know-how & know-why but the time taken to develop all the technology packages are going to be more.

Route C is technological collaboration in selected areas with right partner. Here, we can do indigenous R&D, develop indigenous system but where we face difficulty there we go for limited collaboration with an advanced technology contributor which will make the product world class. This is what we tried in BrahMos.

Route C, the third option is a sure shot and at earliest we are able to realize the best of the product, provided the partnership is perfect and managed well. Having got this experience, the technological collaboration started where the two partners can contribute technologically which means that unless we technologically grow in some form the other partner is not going to support us. Hence, we have gone for a technological collaboration in the case of BrahMos where technological strength of the two countries is being pulled together through two organizations and this Joint Venture has been formed.

The advantages are:

- Less cost for development of new product because of sharing the technology stock and fund contribution from each of the countries.
- Performance characteristics of the new product are unique because we could think of a unique best system, not realized yet.
- Minimum failures due to constant questioning and improvement at every stage.
- Overall reduction in realization time of the deliverable product

However, we had to face many challenges because of our diverse culture, language, expectations and methods of working along with different standards for inspection. All created challenges, but we always looked at maintaining win-win situation throughout the project. In the end, the JV brought out many opportunities like; long standing strategic partnership, core competence in critical technology and to be an international player. The product thus became an international product and opened up export option.

India and Russia are the trusted partners for the last 60 years cooperating in various programmes including some strategic and nuclear programmes and that helped us to formulate the BrahMos Joint Venture. The JV was formed through signing of an Inter Governmental Agreement (IGA) on 12th February 1998 in Moscow with DRDO & NPOM as shareholders with authorized capital of $ 250 million at the ratio of 50.5:49.5. No other JV has got this particular ratio. A Tripartite Agreement was signed between DRDO, NPOM & BrahMos identifying the share of work & process stocks based on the core technological strength. The sharing of work has been done almost equally.

The JV management is also very important and the structure we have adopted is very unique. We have got a board with 6 Directors - 3 members from Russia and 3 members from India and above the board of directors we have formulated a supervisory council which was founded by the two governments. We have also signed with DRDO the total development work and the funds which they have given except some portion. The 95% of the funds to DRDO as PJ-10 programme comes under the control of Comptroller and Auditor General. The government is also happy that there is an auditing system to monitor the Government of India funds, i.e., the control of financial management. BrahMos JV has its own procedure, policies, rules, etc. to enable "Performance". Therefore, we have got flexibility and at the same time control and that's how we have configured this Joint Venture.

Industries have been our partners throughout the journey. We have made industries as a consortium. This JV Company has not made any investment anywhere. All investments are made by industries even before getting the order initially. Today, they are reaping the benefit of it. At present, we are increasing our production capacity because our orders are quite large and as Defence Minister has said earlier, exports are also knocking on our doors. However, at this time we are giving priority to our Indian armed forces. The Government of India is also funding to enlarge the production facilities for supporting the programme. Similarly, the Government of the Russian Federation gave funds for enlarging the production facilities in Russian industries. We are working hand-in-hand. There is no gap and transparency is totally maintained between the two partners.

We made the users and industries as Partners right from the very beginning. We have got Steering Committees with the Services. At every level we have maintained the participation of the user which includes the industry partners and other inspection agencies. Every partner completely becomes part of the system. This is the only project where the inspection agency before the launch confirms the 100 per cent reliability of the missile hitting the target. This is a major factor which has driven this programme.

Throughout the world, there is no answer for the BRAHMOS system, even after 12 years after the BRAHMOS first launch in June 2001. Hence, we are **Number One** in the world. This JV has been developed over the years and the partnership has led to a product which ultimately became Number One. BRAHMOS is a globally competitive product which tops the world in Speed, Precision and Power. It proves that JV helps in realising advanced technology product in a cost-effective manner, produced in shortest possible time giving competitive edge to the nation.

BrahMos enjoys the "Political will" as a national programme. Strong commitment from both countries, joint funding, fairness in partnership, shared vision, high level of commitment, pool of experts available on demand, mutual learning leading to better performance, access to cutting edge technology, long term mutual cooperation, and finally leaping ahead from buyer-seller to equal partners are the contributing factors for the success of BrahMos JV.

This is a win-win partnership for India and Russia. From India's point of view, country's defence potential has increased; has the most cost effective supersonic cruise missile straightaway in the armed forces in shortest time; availability of process stock products and documentations; new technological inputs; key technology to Indian Industries for manufacturing and creation of jobs; equal partnership with Russia and international player through this JV. And from the Russian viewpoint, fund flow and continuation of development activities ensured during the crisis period 1990s; continuous bulk orders for the Russian Industries from India; corporate culture of managing consortium of industries and technological gains with capabilities such as land attack, high precision (Improved Seeker), way points, supersonic steep-dive. Both the countries have gained significantly through this Joint Venture.

In recognizing the great contribution to the strengthening of friendship and cooperation between India and Russia, The Government of Russian Federation has awarded the "Order of Friendship", the highest Russian honour bestowed on foreign citizens, by the President of Russian Federation, Vladimir Putin.

Appendix:

A Primer To Guided Missile

THE STORY OF MISSILE

The basic dictionary explanation for a missile is 'anything that is thrown in the air towards an object is a missile.' Fine! So what could have been the first missile? The rocks hurled by the cave men! But the modern day missiles are nowhere close to the stones. They are much more!

Man has always been looking for newer advancements which have been the reason for success of mankind over other living beings on earth. The cave men were no exception to this tendency. He also wanted to throw big stones accurately and that too for a long distance. He was not able to fair much as the size and distance he was able to throw was directly proportional to his physical strength and the accuracy was dependent on his skill. Then as the civilization grew and man started to understand nature; that understanding is called as science. With his understanding, man was able to build machines like trebuchets. These trebuchets were able to throw rocks for considerable ranges. Some innovative minds hurled Greek fires. These rocks and Greek fires were the missiles of those days.

Then a turnkey event occurred. This happened somewhere around the 10th century. It was the arrival of a mysterious substance. That substance came from China through a very long and treacherous land route called as Silk Route, to Europe. That substance burned heavily when introduced to fire. It was black in colour. The more it was compressed, the more furious it burnt. Since it was black in colour, it was also called as black powder and later as gun powder. This is mechanical, not chemical, compound of sulphur, charcoal and salt-peter (potassium nitrate). Sulphur brought the plasticity to the compound and improves its storage. Salt-peter is the real discovery and available in soil of China and India. This substance was even called the **'Indian snow'** in many countries. Early Chinese used the black powder in fire arrows. This was a transformation in innovation.

World's First Rocket

In the 18th century, a whole new and different dimension was given for the use of Gun powder at the Mysore kingdom by Hyder Ali and later by Tipu Sultan. The invention was that Hyder made Rockets using the gun powder as the charge in a professional manner. Tipu perfected it and produced huge quantities of rockets. He also introduced rocket launchers; each one of which fired three rockets at a time.

A Rocketeers force of about 3,000 men was established by Hyder Ali in 1767 and later Tipu grew the Rocketeers force to more than 5,000 men, with 16 Kushoons (brigades) in each infantry. The Mysore kingdom introduced Rocketeer force with specially trained warriors who could handle these rockets.

Tipu Sultan unleashed the power of rockets to pave way for missile warfare, throughout the world, centuries after his time. What did he do? He effectively utilized the tendency of the gunpowder to burn freely. But when they are enclosed tightly they burn fiercely – the basic principle of crackers. The more tight the enclosure, greater is the explosive force when guided through a nozzle. Tipu exploited these rockets in a professionally designed hardened cast iron cylinder to contain the pressed gun powder. The iron cylinder containment resulted in release of gases, when ignited, escaping from the cylinder with enormous speed through the nozzle. This resulted in the realization of rockets having a reasonable range of travelling more than one km.

Tipu's rockets were of two kinds. One was for the use against the cavalry and the other was used to destroy the ammunition dump of the British. The anti-cavalry rockets had a long sword as weapon attached to the cast iron cylinder by leather strings. When the rockets attained speed, the blades gave stability and when the thrust reduced the blades would rotate in the air. These swirling swords could shred anything in their path into pieces. The next variety resembled the contemporary aerial fireworks. The rockets had long bamboo sticks attached and would fly into the marching enemy soldiers, hit the ground, jump here and there to scare them. Tipu created large scale production set up for manufacturing rockets and its storage in Srirangapatna, his capital fort. The rockets were stored secretly at multiple locations.

The advantages of these rockets are ease of transportability, rapid firing with no recoil and simultaneous deployment in large numbers, creating terrific noise & fire, and swirling with speed. In 1792 Srirangapatna war, Tipu's men launched more than 6000 rockets, with their fearsome speed pierced the British horses. The British cavalries were comfortably defeated and they ran away frightened.

In 1799, in the last war of Mysore, Tipu was killed by the British and they captured 10,000 Indian rockets and sent them to England to their arsenal. The British destroyed all the rocket production plants at Mysore kingdom and brought an end to the serious rocket developers in India, which re-emerged only in 1960s (Ref.: Blazing the Trail – The Early History of Spacecraft and Rocketry by Mike Gruntman, AIAA Publication).

Impressed by the rockets fired at Srirangpatna in 1792 and 1799, a young British inventor William Congreve (later he became Sir) re-engineered Tipu's rockets at the Royal Arsenal, Woolwich. He scaled-up Indian rockets to a range of 3 km, adding more black powder extending the length of the cast iron chamber. He realized that these rockets could become deadly means of war. These re-engineered rockets were later used in naval warfare at various instances. They were used in the Napoleonic Wars including the Battle of Waterloo and later against the Americans including the war at Fort McHenry in September 1814 which created a big fire. The fifth line of national anthem of America reads "And the Rockets' red glare" referring to the use of re-engineered Indian rockets.

Tipu's Rockets are being preserved in the Artillery Museum at Woolwich, England. When I, Mr. NR Iyer, First Project Director of Nag and Mr. Rajat Pal of DRDO had a chance to visit this artillery museum on a Sunday, the curator specially opened the museum for us. I took the rocket on my hand proudly and astonished to see an engineering marvel of Indians who worked with Tipu Sultan to make them. I read the caption at the display saying that "these rockets were used in the 1792 war against the British by Tipu Sultan and the British were defeated". What a proud moment it was for an Indian! Let us now enter into the modern age of missiles.

TERMINOLOGY FOR ROCKETS AND MISSILES

Several types of rockets and missiles are detailed below for ease of understanding. Sounding the rocket is used for upper atmosphere research. These are small rockets with special instrumented payloads which transmits the measurements while flying through the upper atmosphere in its almost vertical trajectory. The launch vehicles of those used for injecting satellites to orbits such as (i) near earth orbit, (ii) polar orbit, (iii) geo synchronous orbit (iv) earth escape orbit for planetary missions. These vehicles are with multiple stages as needed for carrying the satellites or orbiting platforms. The launch vehicles are of two types – Expendable and Re-usable. In expendable launch vehicles, the stages are separated after their performance and fall in their own trajectories. They are not recovered for re-use. In order to reduce the cost of launching, the lower stages are recovered and re-used in the reusable launch vehicles. These are described in Fig 1:

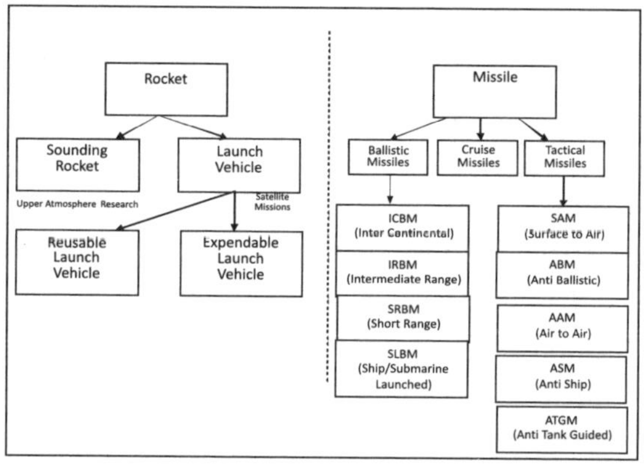

Fig.1: Classification of Rocket and Missiles

The missiles can be classified as Ballistic Missiles, Cruise Missiles and Tactical Missiles. Ballistic Missiles are surface to surface either from land (mobile/fixed/silo) or submarine and in some cases ship. Based on the range they are classified as ICBM, IRBM, SRBM. Tactical missiles are either offensive against specific targets or defensive to protect from the attack. These are also explained in the figure.

Ballistic Missiles

The flight of a ballistic missile can be best understood by throwing a ball to another player, who is standing far. The thrower throws the ball high up in the air with such a velocity that the ball goes high up in the air spending the energy given by the thrower through his throw. It reaches a state where the energy given by the thrower is totally consumed. After losing all these energy, the ball starts falling down to earth due to gravity in a parabolic path and not in a straight path down. If the thrower throws with more energy, the ball goes higher up in the air and lands down farther. Similarly, if he throws with lesser energy, it lands shorter. Thus, the distance the ball covers is dependent on the energy imparted and also on the weight of the ball. For throwing a heavier ball, the thrower has to impart more energy. A ballistic missile works in the above principles. It goes higher and higher based on the multi staging concept and once the energy is exhausted, it flies in the parabolic path towards its destination. During the flight trajectory it has appropriate control and guidance in order to strictly follow the pre-defined path. The ballistic missile is destined to deliver the explosive warhead on the target area. During the return phase of the ballistic missile from a high altitude, it has to re-enter the atmosphere. This is a crucial phase when the temperature of the missile at re-entry reaches 3,000°C. Agni V missile is the right example to understand this. It has a range of 5,000 km. In order to travel this range the missile has to reach an altitude of more than 500 km above the atmosphere and travels back in the ballistic trajectory and re-enters the atmosphere at 85 km altitude. During this re-entry, it travels at a speed 14 times more than the speed of sound and crosses the atmosphere, experiencing a temperature nearing 3,000°C. The shape and the material of the re-entry module need to withstand this condition. Hence, the Agni project started in 1983 as a technology demonstrator, successfully established re-entry technology. India has specialized this vital re-entry technology and today, we have many versions of Agni. The same technology is required for manned space flights, for the re-entry of the Indian astronauts.

Presently, ballistic missiles with multiple nuclear warheads can turn many important cities of enemy into ashes. The cluster like arrangement of warheads called MIRV is propelled out of the atmosphere; the warheads get separated and re-enter the atmosphere individually towards their assigned targets. A typical trajectory of Re-entry mission with MIRV is depicted in Fig. 2.

An important aspect of ballistic missile is that all through its flight path the speed is not constant, rather varies with time at different regimes. This is the factor that differentiates it from cruise missile, which has constant speed.

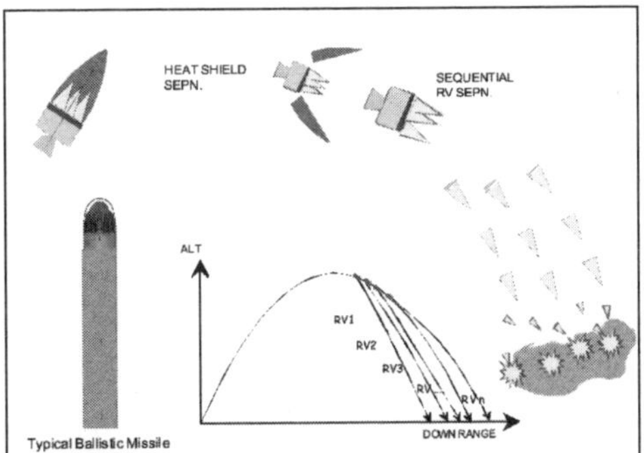

Fig. 2: Typical Trajectory with MIRV

Classification of Missiles

Missiles are generally classified based on their range, type of mission, type of propulsion system and launch configuration. Fig. 3 describes the classification.

Range: Based on the range, ballistic missiles are classified as Short Range Ballistic Missiles (SRBM), Intermediate Range Ballistic Missiles (IRBM), and Inter Continental Ballistic Missile (ICBM). SRBM are called as tactical missile. These missiles reach the distance anywhere less than 1000 km. Generally, they carry a warhead of 500 kg of high energy explosive. Next is the Intermediate Range Ballistic Missiles (IRBM). These missiles reach anywhere up to 5500 km and generally are nuclear capable. Anything that goes beyond this distance is called as Inter Continental Ballistic Missile (ICBM) and will definitively be nuclear capable. After all, one would not like to take the pain of sending a 200 kg explosive through an ICBM and let it drop somewhere on a vast enemy city! The ICBMs can carry nuclear warhead, which can unflinchingly came a major destruction to the enemy's territory.

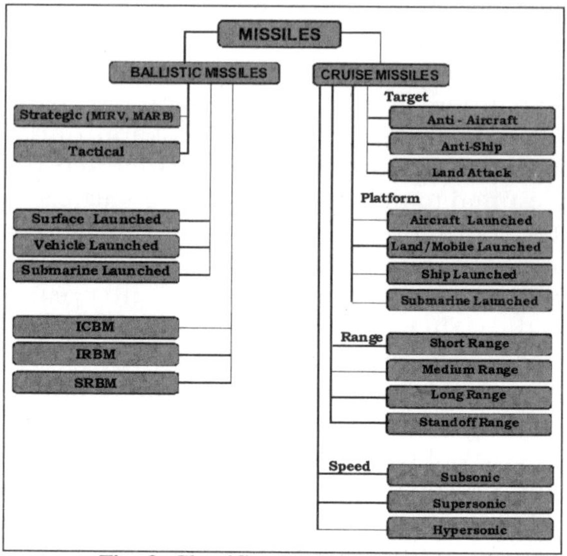

Fig. 3: Classification of Missiles

Propulsion System: Basically there are two types of propulsion schemes: Propulsion systems using liquid propellant (fuel + oxidant) and Propulsion systems using solid propellant (a mix of ammonium perchlorate with binder).

Liquid fuelled missile are generally kept without filling propellant to give long storage life. Once filled the life of the missile becomes limited. Moreover, liquid fuelled missiles need lot of support systems and hence transporting them becomes a horrendous process. Liquid fuels are corrosive and any careless action will end in disaster. Liquid fuelled systems are advantageous in terms of controllability and manoeuvrability but are cumbersome in transportation and fuelling processes.

On the other hand, solid fuel propulsion systems provide exceptional handling capabilities and offer a greater degree of safety than liquid fuelled missiles. Very importantly, they are stored for the determined life and are ready to take off at any time.

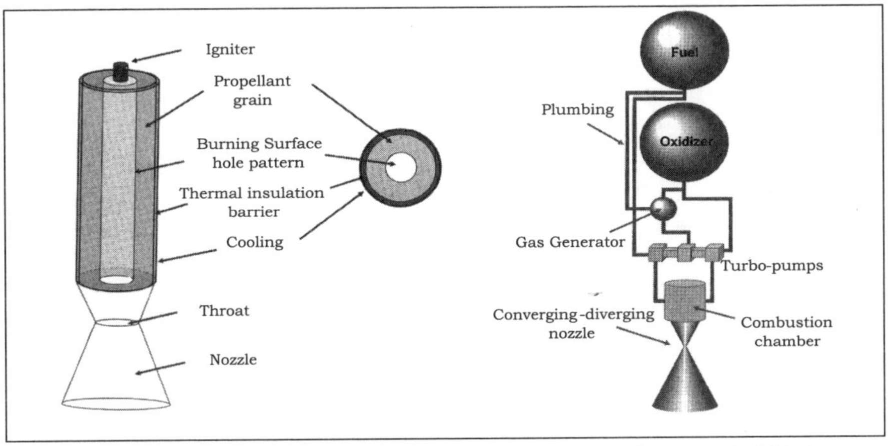

Image Source: Introduction to Rocket Science and Engineering, Travis S Taylor

Fig. 4: Solid and Liquid Propulsion

Launch Configuration: Ballistic missiles can be launched from stationary launch pads or from mobile launchers or from submarines or from underground, i.e., silos.

Stationary ballistic missile launch sites will be the first target for any nation during the initial phase of nuclear war. This is to prevent enemy from launching his own ballistic missiles in retaliation.

Installing the ballistic missiles on submarines is the safest way to protect them from the enemy's first strike. These missiles are called as Submarine Launched Ballistic Missile (SLBM). SLBMs are retaliation weapons and are also known as deterrence weapons. SLBMs are unremitting reminders to the enemy that if they initiate war, they will surely be retaliated.

Ballistic missiles launched from underground silos are similar to the SLBM. Missiles in the silos are not that easy to be destroyed, as they are safe in protected structures.

Another way is through road or rail mobile launchers. The launcher of the missile can also be used for transportation. These are called as Transport Erector and Launcher (TEL) vehicles which can be parked at any desired place to launch the missile. After use, they can quickly be moved away from that place. The purpose of this exercise is to avoid enemy retaliation because by tracking the trajectory, the enemy can trace the location of the missile. The TEL trucks can even be hidden under culverts or bridges and then driven out when needed.

CRUISE MISSILES

Cruise Missiles fly towards a target within the atmosphere, breathing the air like an aircraft. These missiles have wings and other aerodynamic features to fly efficiently and hit a target. The speed with which they fly remains constant and hence these are termed as cruise missiles. Cruise means sailing with a constant speed.

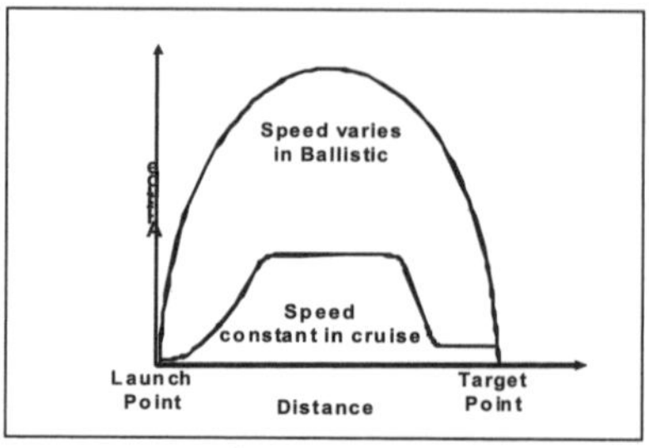

Fig. 5: Trajectories of a Ballistic Missile and a Land Attack Cruise Missile

The trajectories of ballistic missile and land attack cruise missile can be seen in Fig. 5.

Cruise Missiles Classifications

Missiles can be classified on the basis of various orders. The first order is the physical principle of flight which we have seen in the previous section. The second order of classification is the mission of the missile. In other words the intended targets for the missile. After this, the third order is from where the missile is launched? That is the platform that launches the missile.

The typical trajectories that a cruise missile can follow from multiple platforms are depicted in Fig. 6.

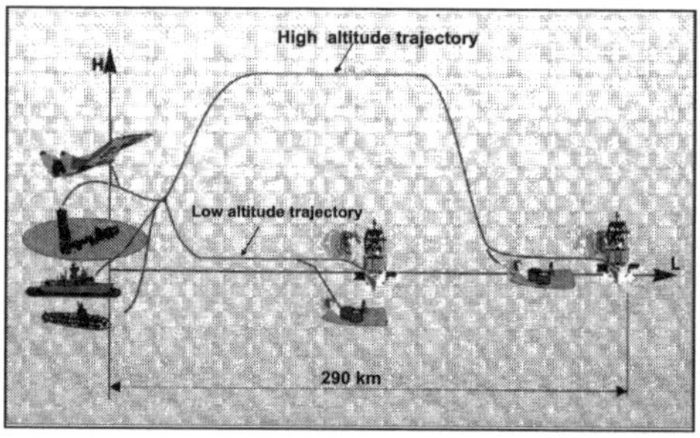

Fig. 6: Trajectories of Cruise Missile

THE TECHNOLOGIES INSIDE A MISSILE

In this section, I will explain what a missile consists of:

What is Inside a Missile?

A missile has the following sections:

- Airframe / Structure
- Propulsion system
- Target Seeker & Guidance and Control system
- Warhead

A brief tabulation is given for understanding the classifications of Cruise Missiles.

II ORDER CLASSIFICATION (MISSION)	**III ORDER CLASSIFICATION (LAUNCH PLATFORM)**
Anti-Ship Cruise Missile (ASCM) Target:All classes of ships, Aircraft Carrier	**Land launched ASCM** (Used For Coastal Defence) Launched From: Surface Based Launchers From Coastal areas
	Ship launched ASCM Launched From: Ships, Submarines
	Air Launched ASCM Launched From: Combat Aircraft
Land Attack Cruise Missile (LACM) Target:Enemy assets on land (Missile base, Production plants, Communication structures, etc.)	**Surface launched LACM** Launched From: Vehicles, Land based fixed launcher
	Ship launched LACM Launched From: Ships & Submarines
	Air launched LACM Launched From: Combat Aircraft

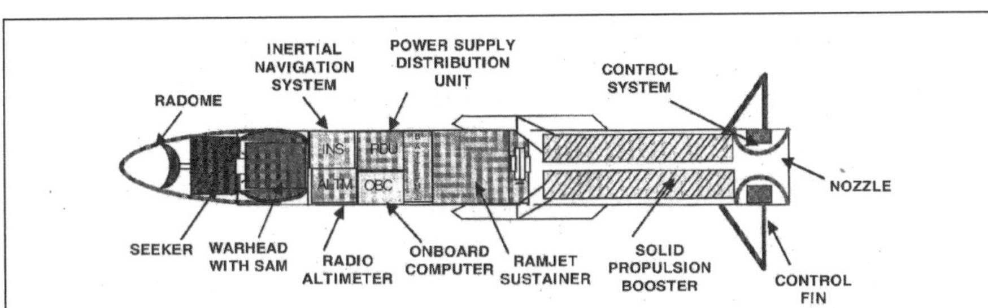

Fig. 7: Inside of a Missile

Fig. 7 shows the various sections of a missile. A ballistic missile will also be similar in structure to the above one, but some of the sections are situated a little differently.

Design

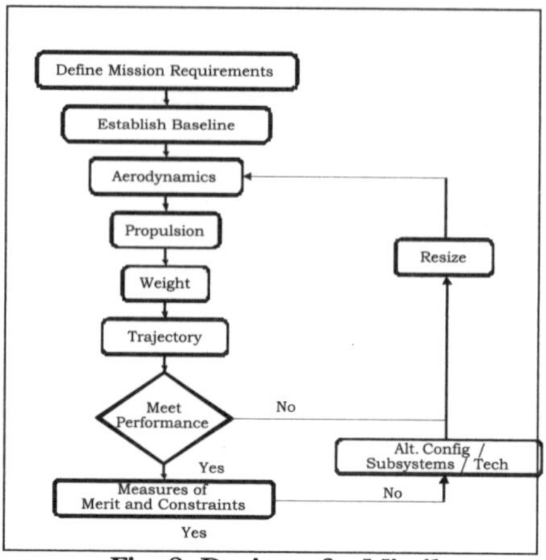

Fig. 8: Design of a Missile

The steps followed in design of typical missile are given in Fig. 8. The missile design process begins with the mission / scenario definition, wherein the basic requirements are specified by the user. Based on this requirement, a baseline is established and then details of aerodynamics, propulsion, structure, control and other features are considered to establish the initial trajectory and performance. Once the feasibility is established, the detailed design work starts and the model testing and wind tunnel testing is progressed parallely.

Airframe

The body of the missile is called as the airframe. The body is aerodynamic in design which enables it to fly faster with a minimum drag. The relation between lift and drag is given in Fig. 9. There are aerodynamic features for controlling the direction of flight. Such features are called as fins, wings and canards based on their size and their placement over the missile. Greater the speed of missile, lesser is the size of features. Their small size is sufficient enough to create the desired aerodynamic effect at high speed. Generally, there are at least any two sets of such features available for a missile, with some rare exceptions. For a ballistic missile the aerodynamic features are less pronounced. Because the ballistic missiles are not required to turn much as in the case of a cruise missile. Moreover, the ballistic missiles are turned through a scheme called as "Reaction Control". The direction of escaping exhaust is controlled to change the direction of the ballistic missile. This is quiet similar to releasing an inflated balloon with the nozzle opened and bent sideways.

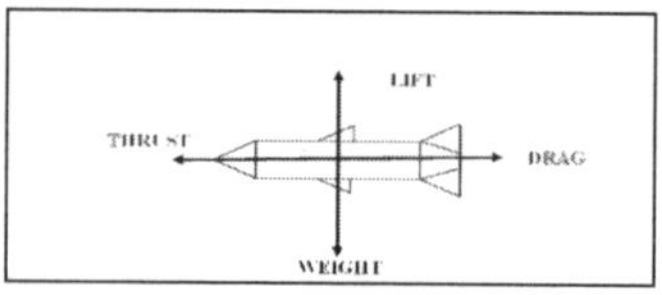

Fig. 9: Lift and Drag Relation

PROPULSION SYSTEM

Propulsion system means the engine of the missile that powers it. Generally depending on the type of the missile, the type of the propulsion also varies. Solid fuel rocket engines and liquid propelled rocket engines are commonly used for ballistic missiles. Solid propellant booster with Ramjet engine for supersonic speed is used in cruise missiles. Turbojet and turbofan are also used for cruise missiles. The efficiency of different propulsion system is calculated based on specific impulse which is defined as the impulse generated per kilogram of fuel. A typical ISP values for different propulsion systems are given in Fig. 10.

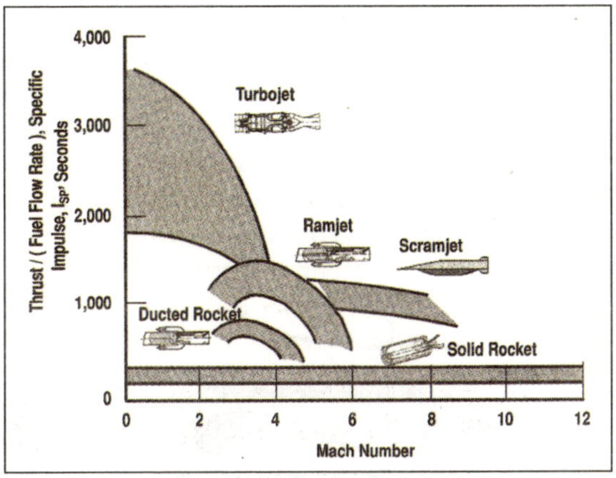

Fig. 10. Specific Impulse for Different Propulsion Systems

Types of Propulsion Systems

Missiles are propelled either by solid or liquid fuel have specific type of engines which are termed as "reaction engine". Missiles fly with reaction engines following Newton's third law, 'for every action there is an equivalent and opposite reaction' in the reaction engines, the fuel is mixed with highly pressurized air containing oxygen (essential for combustion) and burnt. The resultant is that the air expands due to the tremendous amount of heating. The expanding gases escape through a nozzle propelling the missile. But after this scheme of operation the reaction engines can be classified further into two: rocket propulsion and air breathing propulsion.

Rocket Engines

In rocket propulsion systems, an oxidiser and a combusting material (fuel) are burnt in a small cylinder and the resultant hot gases escape through the nozzle. Rocket propulsion systems since they have to carry oxygen can propel only a small payload relative to their weight. But they can travel in the space where there is no oxygen.

Air Breathing Engines

In air breathing propulsion systems, just the fuel is carried and the air in the atmosphere is used to provide oxygen. Air breathing propulsion systems are less heavy and propel greater payloads but can fly only within the atmosphere.

Air breathing engines are classified into pulsejet engines, turbojet engines, turbofan engines and ramjet engines. Air breathing engines collect air through propellers, compress them to pressurize them and then burn the fuel in a combustion chamber. The resulting hot air will escape through the nozzle and during the process they rotate the turbines that are in turn used to rotate the propellers again. So once started, the engine sustains itself. Turbofan engines also follow the same principle with little variation.

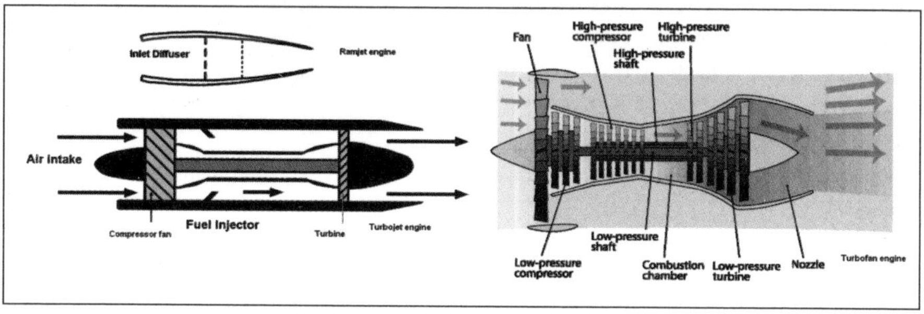

Fig. 11: Types of Air Breathing Engines

In ramjet engines, the air gets compressed automatically due to the supersonic speed of the engine. So, there is no need of compressors but the engine can work only if the air is entering with in supersonic speed. For that, it is essential that the engine must be propelled to supersonic speed through some other propulsion system say a booster rocket.

GUIDANCE & CONTROL SYSTEM

Guidance & Control: You would be surprised to know that the toy 'top' would that you would have played in your childhood plays an important role in guidance. Actually it's derivate called the gyroscope is the key in guidance systems. The guidance system is centred on a computer called as On Board Computer (OBC). This OBC will see that if the missile veers of the course. In what and all ways a missile can veer off? How it will do that? The answer will be very strange.

Since aircraft and missiles are analogous in many ways I choose aircraft for explanation. It is easy since aircraft is a common flying object that we encounter every day.

Assume a flying aircraft. It can undergo displacements along all the three axes (X, Y, Z). These displacements will be straight as the

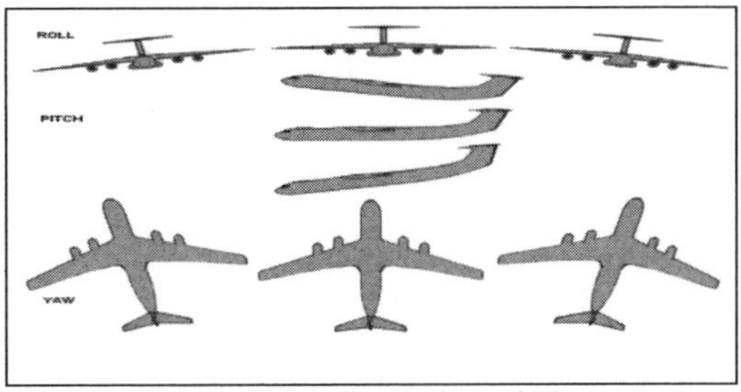

Fig. 12: Roll, Pitch and Yaw

aircraft can move in the fashions of front and back; left and right; up and down. A strong wind from the opposite side of the flight will push the aircraft back and if the wind is flowing from behind, it will push the aircraft along the direction of flight. Similarly, wind blowing from the sides will sway the aircraft sideways. The aircraft can either be lifted or dropped up or down depending upon the wind turbulence.

Resemblance to the three linear displacements, three angular displacements known as Roll, Pitch and Yaw are also possible. Roll, Pitch and Yaw occur along 'X, Y, and Z' axes as axis of rotation respectively.

Pitch occurs when the nose of the aircraft tilted vertically either upwards or downwards. Roll is the angular displacement through which the aircraft rolls to its sides. Yaw is the angular displacement through which the nose of the aircraft is pushed on its sides. The following sketch depicts an aircraft undergoing roll, pitch and yaw.

Though the aircraft is propelled to fly in a straight path but wind and wind induced turbulences can affect the aircraft's heading in six directions by displacing it. To maintain its heading, the aircraft needs to counter these displacements and for this the displacements have to be measured first. Measuring linear displacements is generally done easily whereas gauging the angular displacements is a bit difficult and is quite notorious in altering the course of the aircraft.

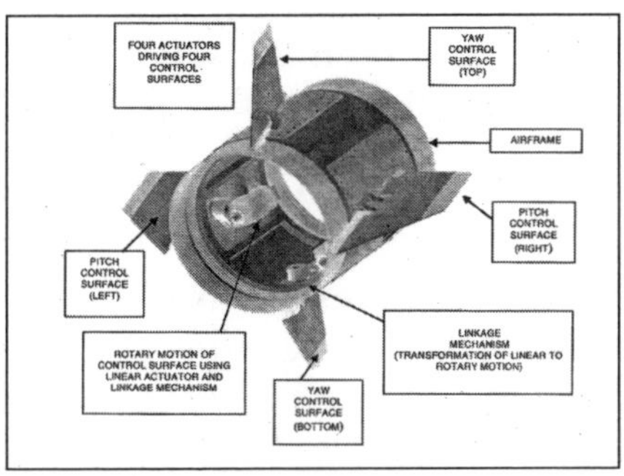

Fig. 13: Typical Control System Schematic

Typical control is explained in the Fig. 3.

If you had noticed the top, it would continue to rotate on its sharp point and will not be able to stand without rotating. As long as it is spinning with a good speed, it continues to stand on its sharp point. You may remember Newton's law of inertia *"A body persists in its state of rest or in a uniform motion unless acted upon by an external unbalanced force"*.

A gyroscope contains a rotor which is supported by movable gimbals. In the figure there are two gimbals. Once, the rotor starts rotating with sufficient speed, the gyroscope stabilizes itself. The roll, pitch and the yaw are supposed to affect the spinning wheel but they are dealt by the movable gimbals. The gimbals move in such a fashion that instead of disturbing the wheel's orientation, the movable frames move accordingly to negate the angular disturbances. There are sensors to measure the movement of the gimbals. The On-Board Computer measures this displacement and actuates the aerodynamic features so that the gimbals moves back to their original position. This is a continuous process which enables the navigation system to maintain the desired course.

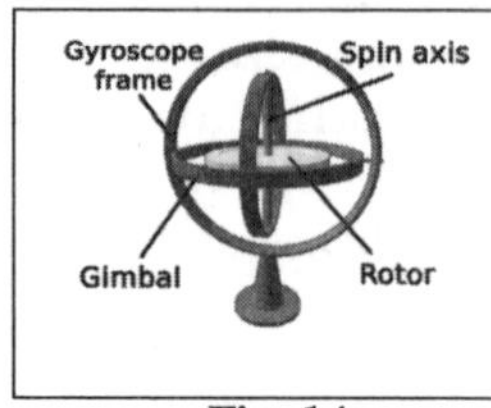

Fig. 14: A Typical Gyroscope

The navigation system, On-Board Processor, Missile Interface Electronics and the Homing seeker form the guidance. Schemes like 'Digital Scene Mapping And Correlation (DSMAC)', 'Terrain Comparison (TERCOM)' etc. are also followed in some of the missiles. These schemes help the missile to find its position relevant to the target and in adjusting to the intended flight path.

TYPES OF GUIDANCE SCHEMES

Guidance systems constantly sense the course that is taken by the missile and then decides the future path by comparing the present and reference course. There are three techniques in determining the course towards the target namely: Homing Guidance System, Command Line of Sight (CLOS) and Navigation guidance system. Based on this technique the guidance systems have been classified.

Command Line of Sight (CLOS)

CLOS guidance systems take inputs regarding the targets locations from the operator directly or indirectly, and decide its course. 'Long Range Surface to Air Missiles' generally have such a type of guidance system. It is sub-classified into Manual-CLOS, Semi Automatic-CLOS (SACLOS) and Automatic-CLOS (ACLOS) systems.

Homing Guidance System

These systems use target a seeker that may detect the target through track laser or radar or Infra-Red rays depending on the type of the seeker used the system is classified so. The next rung of classification is Active and Passive Homing. In active homing the source for the corresponding signals say radar will be in the seeker itself whereas in passive homing some other medium will be illuminating the target.

Navigation Guidance Systems

Navigation guidance systems navigate themselves towards the target by sensing their current course and comparing it with the course to be taken. They know the target's location and their present location from which the course to be taken is determined. The system is quite similar to a man finding the address of a home in a residential area.

These are the most complicated guidance systems since they have to find their course, check it with the reference course and then decide the required course quickly in a very short span of time. Say if a cruise missile is travelling at a speed of 330 m/s, then for every second the missile would have travelled a distance of 330 m. If the guidance system takes one second to calculate its position then before calculating its position the missile would have travelled another 330 m. So the guidance system has to be fast enough so that it is having capability to detect the missile's trajectory and make necessary corrections.

HOMING SEEKER

A target seeker is the eye of the missile, it searches for the target and when found tracks its movement. Guidance system is like the brain of the missile. It takes input from the seeker, locates the target, and calculates that how the missile must move to hit the target. Then it actuates the control system to steer the movements of aerodynamic features to turn the missile. The guidance system constantly takes inputs from the target and keeps on calculating the required trajectory and commands the control system until hit.

There are different types of seekers used in the cruise missile. Radar Seeker is used for anti-ship application. It has small radar that chirps radio frequency energy as pulses and looks for the echoes from the target. The guidance system understands the location of the target from the resultant echoes and steers the missiles towards the target. This seeker is called active radar seeker. Almost all surface to air missiles, anti-ship missiles use this kind of seeker.

When an object is hot, it emits light – whether visible or invisible. Infrared radiations (invisible light) are emitted by a moving aircraft's engines. Some anti-aircraft missiles use an Infrared Seeker composed of tiny infrared sensing elements in an array. These elements detect IR emissions from the engine. Again the guidance system understands the location of the target from the seeker's electronic signals and steer the missile towards the target. There are also laser seekers used, where laser sensing elements like IR sensors are stacked in an array. Many of the anti-tank missiles have laser seekers. There is also Scene Correlation Area Navigation (SCAN) seeker which takes reference from the stored satellite pictures and compares with what it is seeing while nearing the target. These are GPS guided and linked to the On Board Computer.

WARHEAD

Different types of warheads are used for different types of missions. For an anti-tank missile, the warhead must be able to penetrate the thick armour. On the other hand, an anti-aircraft missile requires a warhead that explodes near the target called as continuous rod penetrator warhead. The conventional warheads contain high energy explosive chemicals.

But there are unconventional warheads too. There are three types of unconventional warheads: Nuclear, Biological and Chemical warheads, simply called as NBC. Biological warheads dispense harmful micro organisms like anthrax in the target area; whereas Chemical warheads dispense deadly chemicals like Serene Gas over a target area. Nuclear weapons of today are based on thermo nuclear technology and are much powerful to those dropped on Hiroshima and Nagasaki form an integral part of ballistic missiles. NBC warheads are featured in the LACM varieties also. Generally, these warheads are not armed for detonation before covering certain distance from the launcher. This is called as safe mode and when they approach the target, warhead gets armed through some mechanism. But this may vary depending upon the type of the missile.

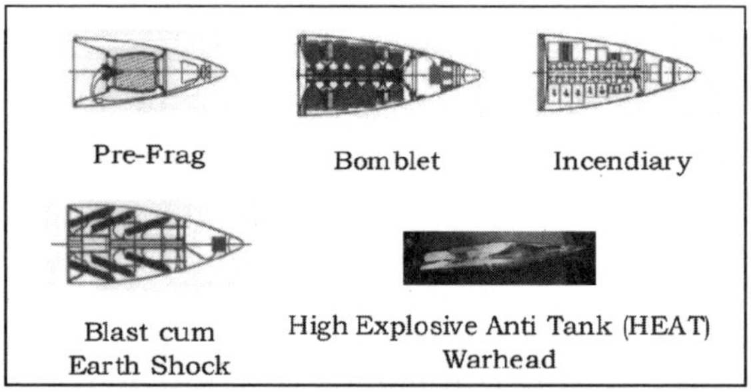

Fig. 15: Different Types of Warheads

For understanding the technicalities present in the book this appendix will help the reader, who may be even from a non-scientific background. For further quest of knowledge internet is the best source. To start with, the reader may refer wikipedia and www.aerospaceweb.org

Bibliography

1. http://treaties.un.org/doc/publication/unts/volume%20299/v299.pdf
2. http://www.nuclearweaponarchive.org/Pakistan/PakOrigin.html
3. Amy Belasco, The Cost of Iraq, Afghanistan, and Other Global War on Terror Operations Since 9/11, March 29, 2011, for Congressional Research Service
4. http://www.flyingbombsandrockets.com/V1_into.html
5. http://csis.org/files/media/csis/pubs/941015lessonsgulfiv-chap03.pdf
6. B Harry Trident; Grandslam and Python: Attacks on Karachi
7. Mike Gruntman, Blazing the Trail
8. SK Ray, Missile Control System, Defence Science Information & Documentation Centre
9. http://kaiser-aeronaut.blogspot.in/2010/03/air-support-at-sea-1971-war.html (A Pakistani perspective)
10. Operation Desert Storm, The Electronic Battle Parts 1 – 3 First published Australian Aviation, June/July/August, 1993, by Carlo Kopp http://www.ausairpower.net/Analysis-ODS-EW.html
11. THE GULF WAR CHRONICLES: A Military History of the First War with Iraq: Richard Lowry Publisher: iUniverse (March 18, 2008) ISBN-10: 1605280062, ISBN-13: 978-1605280066
12. CHAPTER VI - THE AIR CAMPAIGN: www.es.rice.edu/projects/Poli378/ Gulf/gwtxt_ch6.html
13. Gulf War 1991 Chronology1991 Operations after the war http://www.sci.fi/~fta/1991-ops.htm
14. Gulf War Air Power Survey Volume IV Weapons, Tactics, and Training and Space Operations: Publisher: Dept. of the Air Force (January 12, 1994) ISBN-10: 0160429277, ISBN-13: 978-0160429279
15. The Effectiveness of the NATO Tactical Air and Missile Campaign

Against Serbian Air and Ground Forces in Kosovo - A Working Paper: Anthony H. Cordesman Center for Strategic and International Studies: Washington, D.C.

16. World War II: V-1 Flying Bomb, Kennedy Hickman, http://militaryhistory.about.com/od/artillerysiegeweapons/p/v1.htm
17. First strike http://www.sci.fi/~fta/strike-91.htm
18. IRAQ AIR DEFENCE COMMAND: Tom Cooper: December 23, 2001© Orbat.com http://orbat.com/site/agtwopen/IR_ADC.pdf
19. BASIC STRUCTURE OF THE IRAQI INTEGRATED AIR DEFENSE SYSTEM; Appendix VI: http://www.fas.org/man/gao/nsiad97134/app_06.htm
20. CENTRALISED INTEGRATED AIR DEFENCE SYSTEMS and how to construct them in Harpoon 3: Michael Mykytyn: http://www.harpoonhq.com/waypoint/articles/Article_061.pdf
21. http://www.sci.fi/~fta/iraq-9r1.htm
22. http://en.citizendium.org/wiki/Integrated_air_defense_system
23. Dowding of Fighter Command: Victor of the Battle of Britain; Author: Vincent Orange; Publisher: Grub Street (2008) ISBN-13:9781906502140 ISBN-10:1906502145
24. War in the Persian Gulf Almanac: From Operation Desert Storm to Operation Iraqi Freedom; Author: Laurie Collier Hillstrom; Publisher: UXL (2004) ISBN-13: 9780787665630, ISBN-10: 0787665630
25. Desert Storm: A Forgotten War: Author: Archer Jones, Alberto Bin, Richard Hill: Praeger (1998). ISBN-13: 9780275963200, ISBN-10: 0275963209
26. The Iraq Wars and America's Military Revolution: Keith L. Shimko: Cambridge University Press (2010) ISBN-13:9780521111515 ISBN-10: 052111151X
27. The Iraqi Invasion of Kuwait: Religion, Identity and Otherness in the Analysis Otherness in the Analysis of War and Conflict: Hamdi Hassan: Pluto Press (UK) (1999) ISBN-13: 9780745314112 ISBN-10: 0745314112
28. The Encyclopedia of Middle East Wars [5 Volumes]: The United States in the Persian Gulf, Afghanistan, and Iraq Conflicts: Publisher ABC-CLIO (2010) ISBN-13: 9781851099474, ISBN-10:1851099476
29. Technology collaboration: Gupta., S. C, "Growing Rocket Systems and the Team," Prism Books Pvt. Ltd
30. NPOM history: http://www.npomash.ru/history/en/history.htm
31. Tamerlane: Sword of Islam, Conqueror of the World: Justin Marozzi; Da Capo Press (2006) First American Edition (February 27, 2006)

ISBN-10: 030681465X, ISBN-13: 978-0306814655

32. Chapter 9 – Timur's Account of His Invasion of India and Sack of Delhi http://www.ibiblio.org/britishraj/Jackson5/chapter09.html (British Raj. Retrieved 11 December 2012)
33. Alexander The Great: Philip Freeman: Publisher: Simon & Schuster India (2011) ISBN-10: 1416592814, ISBN-13: 978-1416592815
34. Genghis Khan: Life, Death, and Resurrection : Author: John Man: Publisher: St. Martin's Griffin (2007) ISBN: 0312366248 ISBN13: 9780312366247
35. The Mongols (Peoples of Europe): Author: David Morgan: Publisher: Wiley-Blackwell (1991) ISBN-10: 1405135395, ISBN-13: 978-1405135399
36. Genghis Khan And The Making Of The Modern World: Author: Jack Weatherford: Publisher: Crown Publishing Group (2005) ISBN-13: 9780609809648, ISBN-10: 0609809644
37. The Last Mughal: Author: William Dalrymple: Publisher: Vintage (2008) ISBN-10: 1400078334, ISBN-13: 978-1400078332
38. Medieval India: From Sultanat to the Mughals Mughal Empire (15261748) Part Two : Author: Satish Chandra: Publisher: Haranand Publications Pvt Ltd (2006) ISBN-13: 9788124112694 ISBN-10: 812411269X
39. Emperors of the Peacock Throne: The Saga of the Great Mughals; Authored By: Abraham Eraly: Publisher: Penguin (2007) ISBN-13: 9780141001432, ISBN-10: 0141001437
40. THE MUGHAL EMPIRE 1st Edition: Author: Richards: Publisher: Cambridge University Press-new Delhi (2008) ISBN-13: 9788185618494, ISBN-10: 8185618496
41. The Mughal World, Author: Abraham Eraly: Publisher: Penguin Books India (2007) ISBN-13: 9780143102625, ISBN-10: 0143102621
42. Mughal Empire in India (Vol. 3) Revised Edition; Author: S. R. Sharma; Publisher: Atlantic (1999) ISBN-13: 9788171568192 ISBN-10: 817156819X
43. The Last Mughal: The Fall of a Dynasty: Delhi, 1857; Author: William Dalrymple; Publisher: Knopf Doubleday Publishing Group (2007) ISBN-13: 9781400043101, ISBN-10:1400043107
44. Mughal India: Studies In Polity, Ideas, Society, And Culture; Author: M. Athar Ali; Publisher: Oxford University Press (2008) ISBN-13: 9780195696615, ISBN-10: 0195696611
45. Crisis of Empire in Mughal North India: Awadh and the Punjab 1707-48 New edition; Author: Muzaffar Alam; Publisher: Oxford University Press India (2001) ISBN-13: 9780195630008, ISBN-10: 0195630009

46. The Mughal World: India's Tainted Paradise; Author: Abraham Eraly; Publisher: Phoenix House (2008) ISBN-13: 9780753823620 ISBN-10: 0753823624

47. The Mughal Empire; Author: John F. Richards; Publisher: Cambridge University Press (1993) ISBN-13 9780521251198 ISBN-10 0521251192

48. The First Day of the Blitz: September 7, 1940; Author: Peter Stansky; Publisher: Yale University Press (2007) ISBN-13: 9780300125566 ISBN-10: 0300125569

49. The First World War; Authored By: Hew Strachan; Publisher: Penguin (2005) ISBN-13: 9780143035183, ISBN-10: 0143035185

50. The First World War, Second Edition: A Complete History; Author: Martin Gilbert; Publisher: Holt McDougal (2004) ISBN-13: 9780805076172, ISBN-10: 0805076174

51. Fighting in Hell: The German Ordeal on the Eastern Front; Publisher: Random House Publishing Group (1997) ISBN-13: 9780804116985 ISBN-10: 0804116989

52. The Third Reich at War, 1939-1945 ; Author: Richard J. Evans; Publisher: Penguin Books (2010) ISBN-13: 9780143116714 ISBN-10: 0143116711

53. Leningrad 1941-44: The Epic Siege; Author: Robert Forczyk; Publisher: Osprey Publishing (UK) (2009) ISBN-13: 9781846034411, ISBN-10: 1846034418

54. Leningrad Under Siege: First-Hand Accounts of the Ordeal; Author: Ales' Adamovich, Daniil Granin; Publisher: Pen & Sword Books (2008) ISBN-13:9781844154586, ISBN-10: 1844154580

55. GDP values: Harrison, Mark, "The Economics of World War II: Six Great Powers in International Comparison", Cambridge University Press (1998). ISBN-10: 0521785030, ISBN-13: 978-0521785037

56. Milward, Alan S., "War, economy, and society, 1939-1945", University of California Press (1979). ISBN-10: 0520039424 ISBN-13: 978-0520039421

57. Overy, Richard, "Why the Allies Won", W. W. Norton & Company; Reprint edition (1997). ISBN-10: 039331619X, ISBN-13: 978-0393316193

58. Barnett, Correlli, "The audit of war : the illusion & reality of Britain as a great nation", Pan Books; First Edition (June 2001) ISBN-10: 033034790X, ISBN-13: 978-0330347907

59. Weapons and manpower: http://www.historylearningsite.co.uk/weapons_and_manpower.htm

60. Mark Harrison: Industrial mobilization for World War II: a German

comparison:http://www2.warwick.ac.uk/fac/soc/economics/staff/academic/harrison/public/opk2000mobilisation.pdf

61. Thoughts for Change – We Can Do It; APJ Abdul Kalam, A Sivathanu Pillai, Pentagon Press New Delhi, 2013
62. БОЕВОЙ И ЧИСЛЕННЫЙ СОСТАВ ВООРУЖЕННЫХ СИЛ СССР В ПЕРИОД ВЕЛИКОЙ ОТЕЧЕСТВЕННОЙ ВОЙНЫ (1941-1945 гг.): Статистический сборник № 1 (22 июня 1941 г.) [Battle and Strength of USSR Armed Forces Between World War II (1941-1945 gg.) Statistical collection number 1(22 June 1941)]: http://www.teatrskazka.com/Raznoe/BiChSostavVS/BiChSostavVS.html
63. The Lion and the Tiger: The Rise and Fall of the British Raj 1600-1947: Denis Judd: Oxford University Press, USA (2010) ISBN-13 9780192805799 ISBN-10 0192805797
64. Raj : The Making And Unmaking Of British India: Lawrence James: Little, Brown Book Group (1998) ISBN-13: 9780349110127 ISBN-10: 0349110123
65. The British Raj In India: History Map Series: Diarmuid Jeffreys: New Holland Publishers UK (2008) ISBN-13: 9781845377588, ISBN-10: 1845377583
66. Wavell and the Dying Days of the Raj: Britain's Penultimate Viceroy in India: Muhammad Iqbal Chawla:: Oxford University Press (2012) ISBN-13: 9780199062751, ISBN-10: 0199062757
67. Confronting Colonialism: Resistance and Modernization Under Haidar Ali & Tipu Sultan: Irfan Habib (Editor): Anthem Press (2002) ISBN-13: 9781843310242, ISBN-10: 1843310244
68. The Dreams of Tipu Sultan And Bali 3rd Edition: Girish Karnad: Oxford University Press (2004) ISBN-13: 9780195664768 ISBN-10: 0195664760
69. HAIDAR ALI & TIPU SULTAN-HB 01 Edition: PRAKASH: Anmol Publications Pvt.Ltd. (2002) ISBN-13: 9788126110858 ISBN-10: 8126110856
70. History of Tipu Sultan: Mohibbul Hasan: AAKAR BOOKS (2009) ISBN-13: 9788187879572, ISBN-10: 8187879572
71. Tipu Sultan (1750-1799): S. Ram Shiv Gajrani: Publisher: Commenwealth Publishers (2011) ISBN-13: 9788131103180, ISBN-10: 8131103188
72. Tipu Sultan (A Life History): Md. Ilyas Nadvi: Genuine Publications (2011) ISBN-13: 9788189965235, ISBN-10: 8189965239
73. Michael J. Lyons: World War II: A Short History: Prentice Hall (2009) ISBN-13:9780205660568, ISBN-10: 0205660568

74. H. P. Willmott, Charles Messenger, Robin Cross: World War II Dk Publishing (Dorling Kindersley) (2004) ISBN-13: 9780756605216 ISBN-10: 0756605210

75. Elizabeth Raum, William O Oldson: World War II: You Choose Books ISBN: 1429623446 EAN: 9781429623445

76. Sterling Michael Pavelec (editor): The Military-Industrial Complex and American Society: ABC-CLIO January 2010 ISBN:978-1-59884-187- e, ISBN: 978-1-59884-188-6

77. Paul Koistinen: 'Military-industrial Complex: A Historical Perspective': Publisher: Praeger Publishers Inc (November 1980) ISBN-10: 0030557666, ISBN-13: 978-0030557668

78. Carroll W. Pursell: Military-industrial Complex: Joanna Cotler Books (February 22, 1973) ISBN-10: 006045296X
ISBN-13:978-0060452964

79. David Christopher Naylor Swanson (editor): David Swanson (December 20, 2011): The Military Industrial Complex at 50 ISBN-10: 098308307X, ISBN-13: 978-0983083078

80. Masako Ikegami-Andersson: 'The Military-Industrial Complex: The Cases of Sweden and Japan': Dartmouth Pub Co (March 1993) ISBN-10: 1855213095, ISBN-13: 978-1855213098

81. Sidney Lens: 'The Military-Industrial Complex': Pilgrim Press; 1st edition (1970) ISBN-10: 0829801545, ISBN-13: 978-0829801545

82. Steven Rosen: Testing the theory of the military-industrial complex: Lexington Books (1973) ISBN-10: 0669848719, ISBN-13: 978-0669848717

83. Senator William Proxmire: 'Report from Wasteland: America's Military-Industrial Complex': Praeger; First Edition (1970) ISBN-10: 0275026507, ISBN-13: 978-0275026509

84. Omer L Carey: The Military-Industrial Complex and United States Foreign Policy: Washington State University Press (1969) ASIN: B001AG4U7C

85. Mikhail Agursky: The Soviet military-industrial complex (Jerusalem papers on peace problems): Magnes Press, Hebrew University (1980): ASIN: B0006E5E4C

Abbreviations

A&N	-	Andaman and Nicobar
AAA	-	Anti-Aircraft Artillery
AAR	-	After Action Report
ACLOS	-	Automatic Command Line of Sight
ADDG	-	Aerospace Design and Dynamics Group
ADOC	-	Air Defence Operations Centre
ADS	-	Air Defence Systems
ANURAG	-	Advanced Numerical Research and Analysis Group
AOR	-	Area Of Responsibility
ASCM	-	Anti Ship Cruise Missile
ATM	-	Anti Tank Missile
ATVP	-	Advance Technology Vehicle Programme
AWACS	-	Airborne Warning And Control Systems
BDA	-	Battle Damage Assessment
BDL	-	Bharat Dynamics Limited
BMD	-	Ballistic Missile Defence
BWC	-	Biological Weapons Convention
C2	-	Command and Control
C3	-	Command, Communication and Control
CCTV	-	Closed Circuit Tele Vision
CDC	-	Combat Direction Centre
CFD	-	Computational Fluid Dynamics
CIS	-	Commonwealth of Independent States
CLOS	-	Command Line of Sight
CMS	-	Combat Management System
CNES	-	Centre Nationale de Etudes Spatiales

CPHS	-	Combined Pneumo Hydraulic System
CT	-	Computerized Tomographic
CTBT	-	Comprehensive Test Ban Treaty
CWC	-	Chemical Weapon Convention
DAE	-	Department of Atomic Energy
DCNS	-	Deputy Chief of Naval Staff
DRDL	-	Defence Research and Development Laboratory
DRDO	-	Defence Research and Development Organisation
DSMAC	-	Digital Scene Mapping And Correlation
ECCM	-	Electronics Counter Counter Measures
ECM	-	Electronic Counter Measure
EMCON	-	Emission Control
ESM	-	Electronic Support Measure
FAC	-	Fast Attack Craft
FCC	-	Fixed Command Centre
FCS	-	Fire Control System
FOC-in-C	-	Flag Officer Commanding-in-Chief
FOG	-	Fibre Optic Gyro
FRR	-	Flight Readiness Review
G&B	-	Godrej & Boyce
GMDP	-	Guided Missile Development Programme
GSLV	-	Geosyncronise Satellite Launch Vehicle
GSQR	-	General Staff Qualitative Requirement
GUI	-	Graphical User Interface
HAL	-	Hindustan Aeronautics Limited
HILS	-	Hardware In Loop Simulation
HPCL	-	Hindustan Petroleum Corporation Limited
IADS	-	Integrated Air Defence System
IAF	-	Indian Air Force
IGA	-	Inter Governmental Agreement
ICBM	-	Inter Continental Ballistic Missile
IGCAR	-	Indira Gandhi Centre for Atomic Research
IGMDP	-	Integrated Guided Missile Development Programme
IISc	-	Indian Institute of Science
ILTP	-	Integrated Long Term Programme
IMU	-	Inertial Measurement Unit

INS	-	Inertial Navigation System
IOC	-	Intercept Operations Centre
IOCL	-	Indian Oil Corporation Limited
IRBM	-	Intermediate Range Ballistic Missiles
ISP	-	Specific Impulse
ISRO	-	Indian Space Research Organisation
ITR	-	Integrated Test Range
L&T	-	Larsen & Toubro
LACM	-	Land Attack Cruise Missile
LCS	-	Launcher Control System
MAI	-	Moscow Aviation Institute
MAL	-	Mobile Autonomous Launcher
MANPADS	-	Man Portable Air Defence System
MCP	-	Mobile Command Post
MIC	-	Military Industry Complex
MIRV	-	Multiple Independent Re-entry Vehicle
MMI	-	Man Machine Interface
MNC	-	Multi National Company
MOU	-	Memorandum of Understanding
MRR	-	Mission Readiness Review
MRV	-	Mobile Replenishment Vehicle
MST	-	Missile Study Team
MTCR	-	Missile Technology Control Regime
NASA	-	National Aeronautics and Space Administration
NBC	-	Nuclear, Biological and Chemical
NCW	-	Network Centric Warfare
NID	-	National Institute of Design
NPOM	-	NPO Mashinostroyenia
NPT	-	Non-Proliferation Treaty
OBC	-	On Board Computer
OEM	-	Original Equipment Manufacturers
ONGC	-	Oil and Natural Gas Corporation
PAC	-	Public Accounts Committee
PCD	-	Peripheral Control Device
PFFR	-	Pokhran Field Firing Range
PIV	-	Particle Image Velocimetry
PLIF	-	Planar Laser Induced Fluorescence

PSLV	-	Polar Satellite Launch Vehicle
PSU	-	Public Sector Undertakings
PXE	-	Proof and Experimental Establishment
QR	-	Qualitative Requirements
RATO	-	Rocket Assisted Take Off
RCS	-	Radar Cross Section
REX	-	Re-Entry Experiment
RFID	-	Radio Frequency Identification
RFNA	-	Red Fuming Nitric Acid
RFP	-	Request For Proposal
RLG	-	Ring Laser Gyro
SA to RM	-	Scientific Advisor to Raksha Mantri (Defence Minister)
SACLOS	-	Semi Automatic-CLOS
SAIL	-	Steel Authority of India Limited
SAM	-	Surface to Air Missile
SCAN	-	Scene Correlation and Area Navigation
SCRAMJET	-	Supersonic Combustion Ramjet
SEP	-	Société Européenne de Propulsion
SLBM	-	Submarine Launched Ballistic Missile
SLV-3	-	Satellite Launch Vehicle-3
SOC	-	Sector Operations Centre
SRBM	-	Short Range Ballistic Missile
SWDT	-	Special Weapons Development Team
TEL	-	Transport Erector and Launcher
TERCOM	-	Terrain Comparison
TNT	-	Tri-Nitro-Toluene
TVC	-	Thrust Vectored Control
USAF	-	United States Air Force
UVLM	-	Universal Vertical Launcher Module
VSSC	-	Vikram Sarabhai Space Centre
WMD	-	Weapons of Mass Destruction
WW-I	-	World War I
WW-II	-	World War II